▼ Adaptation and Natural Selection in Caves

▼ Adaptation and Natural Selection in Caves

The Evolution of *Gammarus minus*

DAVID C. CULVER
THOMAS C. KANE
DANIEL W. FONG

HARVARD UNIVERSITY PRESS
Cambridge, Massachusetts
London, England
1995

Copyright © 1995 by the President and Fellows of Harvard College
All rights reserved
Printed in the United States of America

This book is printed on acid-free paper, and its binding materials have been chosen for strength and durability.

Library of Congress Cataloging-in-Publication Data

Culver, David C., 1944–
    Adaptation and natural selection in caves : the evolution of Gammarus minus / David C. Culver, Thomas C. Kane, Daniel W. Fong.
       p.   cm.
    Includes bibliographical references (p.   ) and index.
    ISBN 0-674-00425-6 (acid-free paper)
    1. Gammarus minus—Evolution. 2. Natural selection. 3. Cave fauna. I. Kane, Thomas C. II. Fong, Daniel W. III. Title.
QL444.M315C84 1995
595.3'71—dc20   94-5362
                CIP

Dedicated to

Thomas C. Barr, Jr.
Kenneth A. Christiansen
Thomas L. Poulson

—*who blazed a trail and trained us to follow*

▼ Acknowledgments

We have been extraordinarily fortunate in sharing in the collegiality of cave biologists throughout the world, and in having been given the assistance of numerous cavers and the permission of countless landowners to enter caves on their property. Rather than risk thoughtless omissions, we thank all three groups collectively.

We owe a particularly heavy debt of gratitude to three colleagues without whose help this book would not have been possible. John R. Holsinger of Old Dominion University introduced one of us (DCC) to the *Gammarus minus* "problem" nearly thirty years ago and has provided a much-needed systematist's perspective to our work. William K. Jones of the Karst Waters Institute did most of the hydrological work in our study areas and accompanied us on numerous collecting trips. Robert W. Jernigan of American University provided the statistical expertise, combined with a real feel for the data, that allowed us to summarize much of our work in a particularly elegant way.

We are also grateful to the following individuals for making comments on earlier drafts of this book: John R. Holsinger, David P. Mindell, Thomas L. Poulson, and Robert C. Richardson.

The National Science Foundation provided support for most of the work described here.

Finally, we thank our families for their forbearance throughout the writing of the book.

Contents

1. Introduction   1
2. Caves as Evolutionary Laboratories   4
3. *Gammarus minus* as a Model Organism   32
4. The Ecological Theater   48
5. The Geography of *Gammarus minus*   69
6. Making a Case for Selection   118
7. Putting the Pieces Together   149
8. Questions of Time   171
9. Adaptation in *Gammarus minus*   187

Glossary   197

References   203

Index   219

▼ Adaptation and
Natural Selection
in Caves

1 Introduction

More than 135 years after the publication of *On the Origin of Species,* Darwin's primary argument of evolution by natural selection remains largely untested. This is to us the fundamental problem in evolutionary biology—distinguishing patterns that are a consequence of common descent from patterns that are the result of natural selection. For example, two butterfly species may have similar brightly colored wing patterns because they share a common ancestor with the distinctive wing pattern, or because each species independently evolved the pattern as warning coloration to predators. These wing patterns may inform us about phylogenetic relationships among species or they may inform us about the selective pressures and adaptive responses of species, but they will not inform us about both. The teasing apart of historical patterns, typically the province of systematics, and adaptive patterns, typically the province of evolutionary ecology, continues to engage evolutionary biologists.

Darwin (1859) suggested that classification should reflect "propinquity of descent." After devoting nearly all of *On the Origin of Species* to making a case for the primacy of natural selection as the driving force of evolution, Darwin, in the penultimate chapter, recognized that adaptive characters, the very products of natural selection, were among the least valuable in establishing a classification. He noted that adaptive features, such as the body shapes of whales and fish, were often analogous characters, which are the result of adaptation to similar environments and not of common ancestry. Systematists continue to develop techniques and approaches for revealing "true phylogenies" (that is, classification schemes that reflect patterns of descent). Cladistics, an approach utilizing shared derived character states rather than ancestral character states in phylogeny construction, is a particularly relevant example.

Modern evolutionary biologists are assessing the pervasiveness of adaptation. The neo-Darwinian synthesis and the rise of population genetics have provided rigorous alternatives to natural selection as explanations for evolutionary phenomena; an example is the neutral theory of molecular evolution (Kimura 1983). Selectionist explanations have often been incomplete (Brandon 1990) or have been accepted uncritically (Levins and Lewontin 1985). Recently much emphasis has been placed on measuring selection in nature (Lande and Arnold 1983, Endler 1986) in order to demonstrate clearly a relationship between a presumptively adaptive feature and the fitness of its possessor.

In this book we examine the role of evolutionary history and natural selection in molding the morphology of the cave-inhabiting crustacean *Gammarus minus*. Caves have long intrigued evolutionary biologists because, given the scarcity of light and food in caves, they have been viewed as harsh and unusual environments. Evolutionists have often viewed cave animals as bizarre as well, particularly given the common absence of eyes and pigmentation that characterizes their striking morphology. Thus, although evolutionary biology has by no means ignored caves and cave animals, it has often considered them to be oddities with little to offer the main body of evolutionary theory.

We argue that caves and cave animals are valuable empirical models for the study of evolution, particularly for the study of adaptation. Specifically, cave animals offer much evidence on the issue of evolutionary tradeoffs, a recurring theme in the study of adaptation. The cave environment, less complex than other types of habitats, is an advantage because selection pressures can be more clearly recognized in a simpler context. The "bizarre" morphology of cave animals is also an advantage because some highly distinctive features are repeated, both among disparate taxa from the same cave system and among related taxa isolated in geographically distant cave systems.

Highly discontinuous systems in which habitats and morphological types are repeated (for example, Galápagos finches and Hawaiian *Drosophila*) have often been the setting for classic evolutionary studies. We believe caves offer many of the same advantages and much of the same potential.

We also suggest that the morphology of cave animals has been misrepresented and misinterpreted both by specialists in cave biology and by evolutionary biologists in general. An overemphasis on lost features, particularly eyes and pigment, has made cave-dwellers appear bizarre, and a lack of emphasis on elaborated features, particularly extra-optic sensory structures, has diminished their value as evolutionary models. A holistic view of the morphology indicates that cave animals are quintessential examples of the consequences of evolutionary tradeoffs.

We will treat *Gammarus minus* and the *G. minus* model at several levels. This book is, first and foremost, a detailed and multifaceted evolutionary study of a specific organism in a particular environmental and geographic setting. We detail the biology of the organism with regard to the geographic setting in which it exists and the environmental conditions to which it is exposed. At another level, the *G. minus* model we present here is a paradigm for cave colonization and adaptation. We compare our data with theoretical models of speciation and adaptation in caves and with empirical studies of other species of cave animals. Finally, in the broadest context, we view the *G. minus* model as a general case study of the role of natural selection and adaptation in evolution.

2 Caves as Evolutionary Laboratories

Dragon Larvae

The first organism to be described as a cave-dweller was the salamander *Proteus anguinus*. An inhabitant of some of the large caves in the classic Slovenian karst of eastern Europe, *P. anguinus* was included in J. N. Laurenti's 1768 *Synopsis Reptilium* and, like many cave organisms identified later, was noted for its curious appearance. Periodically the salamanders, which are colorless and which retain some juvenile features in adulthood, would be washed out of the caves into large, enclosed depressions called "poljes." Thus exposed, they were believed by the people of the region to be the larvae of a type of dragon that provoked floods by its movements (Bellés 1992).

Today thousands of cave-limited species are known, and there are probably between 50,000 and 100,000 obligate cave-dwelling species (Culver and Holsinger 1992). The often bizarre morphology of cave organisms, together with the exotic nature of their habitat, has long captured the interest of evolutionary biologists. Although certainly better known than most cave-dwelling species, *Proteus* exemplifies many of the characteristics that fascinate biologists.

The vast majority of obligate cave animals are blind and depigmented (Vandel 1964). In the case of *Proteus* there is the rather unusual complication that the larvae have small eyes that degenerate during development, leaving the adults functionally blind. *Proteus* has other bizarre characteristics as well. Its body is, in general, more slender, and its appendages are relatively longer than those of most salamanders (Fig. 2.1). This "fragility" of appearance and the elaboration

Fig. 2.1. The European cave salamander *Proteus anguinus*. Length from snout to vent at maturity is approximately 20 cm. From Botosaneanu (1986).

of extra-optic sensory structures (for example, lateral lines in fish, antennae in arthropods) are also characteristic of most cave-limited species (Culver 1982).

With no close surface-dwelling relatives, *Proteus* has long been placed in the family Necturidae with only one other genus—the North American mud puppies *Necturus*—or in a family by itself. Many cave-limited species are in genera and families that are exclusively subterranean, and many others are in genera in which surface-dwelling species are geographically distant.

Proteus has a highly restricted geographic range, limited to a portion of Italy and Slovenia. In fact, its range is larger than that of most cave-limited species. Many cave-limited species occupy only a single cave or group of adjoining caves. Of the 160 cave-limited species described from Virginia and West Virginia caves, 55 are known from a single cave (Culver and Holsinger 1992).

The life cycle of *Proteus* is highly modified relative to that of surface-dwelling salamanders: it is characterized by greater longevity, lower reproductive rate, and neoteny. Low reproductive rate and great longevity are hallmarks of other cave salamanders and of cave-limited species in general (Culver 1982). For example, in an intensive study of the crayfish *Orconectes australis australis* in Shelta Cave, Alabama, Cooper (1975) observed only three egg-bearing females in a population of over 400 crayfish over a six-year period.

Cave species thus appear to be a study in extremes—in their morphology, in their lack of obvious relatives, in their distribution, and in their life cycle. What have evolutionary biologists made of these creatures? What use have they made of the natural laboratories presented by caves?

Early Evolutionary Theorists

Lamarck

The term *Lamarckism* conventionally refers to the evolution of acquired characters even though Lamarck himself considered it a secondary mechanism in his more complex theory (Kane and Richardson 1985). The eyelessness and lack of pigment of many cave animals would seem to be a Lamarckian dream come true. What better example could there be of the theory of use and disuse than an organism that has lost its sight and its color in a dark environment? Curiously, although Lamarck knew of *Proteus,* he apparently attached little importance to it. It makes only one appearance in his most famous treatise, *Zoological Philosophy:*

> The *Proteus,* an aquatic reptile allied to the salamanders, and living in deep dark caves under the water, has, like *Spalax,* only vestiges of the organ of sight, vestiges which are covered up and hidden in the same way ... Light does not penetrate everywhere; consequently animals which habitually live in places where it does not penetrate have no opportunity of exercising their organ of sight, if nature has

endowed them with one. Now animals belonging to a plan of organization of which eyes were a necessary part, must have originally had them. Since, however, there are found among them some which have lost the use of this organ and which show nothing more than hidden and covered up vestiges of them, it becomes clear that the shrinkage and even disappearance of the organ in question are the results of permanent disuse of that organ. (Lamarck 1984, p. 116)

To Lamarck, cave animals were simply eyeless, as were soil-dwelling mole-rats. He either ignored or was unaware of the elaborated characteristics of cave animals.

Darwin

While caves and cave faunas were little understood in Lamarck's time, additional information was available by the mid-nineteenth century, at the time Darwin was formulating his ideas. Extensive surveys of cave faunas of the Balkan peninsula had begun in the 1830s, and numerous species had been described by 1850. One of the cave fish from Mammoth Cave, Kentucky, *Amblyopsis spelaea*, was described by J. E. De Kay in 1842. Darwin mentions cave animals in all editions of *On the Origin of Species*. Unfortunately, his knowledge was limited and based on some erroneous information. For example, he accepted Benjamin Silliman's report that a cave rat, *Neotoma*, collected in Mammoth Cave, initially had large eyes but was functionally blind and that, "after having been exposed for about a month to a graduated light, [*Neotoma*] acquired a dim perception of objects" (Darwin 1872, pp. 61–62). These unsubstantiated and erroneous observations of Silliman (*Neotoma* is not functionally blind) led Darwin to view the loss of eyes in cave animals as the result of disuse.

To Darwin as to Lamarck, cave animals were examples of eyelessness and of loss of structure in general. For Darwin the explanation for eye loss was rather simple and straightforward. As he argued consistently from the first through the sixth edition of the *Origin of Species:* "It is well known that several animals, which inhabit the caves of Carniola and Kentucky, are blind . . . As it is difficult to imagine that eyes, though useless, could be in any way injurious to animals living in darkness, their loss may be attributed to disuse" (ibid., p. 135).

To explain other cases where there were clear losses of morphological structure, Darwin did not hesitate to invoke natural selection. For example, he argued that winglessness of island-dwelling beetles might be the consequence of the dangers associated with flight in these habitats. Indeed, one could use his arguments, as did the neo-Darwinians discussed below, concerning reduction of the head in the parasitic barnacle *Proteolepas* nearly verbatim to explain the loss of eyes and pigment in cave animals: "Now the saving of large and complex structure, when rendered superfluous, would be a decided advantage to each successive individual of the species; for in the struggle for life to which every animal is exposed, each would have a better chance of supporting itself, by less nutriment being wasted" (ibid., p. 143).

Darwin was interested in the evolutionary relationships of cave animals to surface-dwelling animals. He looked at convergent morphologies among cave animals not as evidence for natural selection but rather as a mask obscuring evolutionary relationships. He used the evolutionary relationships to attack creationist views, rather than to argue for the primacy or importance of natural selection:

> It is difficult to imagine conditions of life more similar than deep limestone caves under a nearly similar climate; so that on the common view of the blind animals having been separately created for the American and European caverns, close similarity in their organization and affinities might have been expected; but, as Schiodte and others have remarked, this is not the case, and the cave-insects of the two continents are not more closely allied than might have been anticipated from the general resemblance of the other inhabitants of North America and Europe. (Ibid., p. 138)

Darwin also addressed the disjunct distributions of many cave animals and the fact that many cave species appeared to lack close surface-dwelling relatives:

> Far from feeling surprise that some of the cave-animals should be very anomalous, as Agassiz has remarked in regard to the blind fish,

the *Amblyopsis,* and as is the case with blind *Proteus* with reference to the reptiles of Europe, I am only surprised that more wrecks of ancient life have not been preserved, owing to the less severe competition to which the scanty inhabitants of these dark abodes will have been exposed. (Ibid., pp. 136–137)

Darwin's statement is in part, undoubtedly, a refutation of the creationist views of Agassiz. However, it also argues that many cave species are relicts and that the cave habitat provides something of a museum for their preservation. The relictual nature of cave species has continued as a theme of orthogenetic (Vandel 1964) and neo-Darwinian theories (Barr 1968).

Interesting though his views are, especially in the way they presage modern controversies, Darwin did not address the central question of the evolution of cave animals—why are they similar in both their reduced and elaborated characters? In his discussion of cave animals, largely devoted to disuse and their evolutionary connection to surface-dwelling ancestors, he mentions natural selection only in passing: "natural selection will often have effected other changes, such as an increase in the length of antennae or palpi, as a compensation for blindness" (ibid., p. 136). Employing a pattern of reasoning that was common among naturalists of the nineteenth century, Darwin explained the evolution of cave animals in four steps (Kane and Richardson 1985):

1. IMMIGRATION: migration of surface forms into the cave habitat
2. DEGENERATION: atrophy and underdevelopment of visual organs because of disuse
3. INHERITANCE: transmission of environmentally induced reductions to offspring
4. COMPENSATION: hypertrophy of extra-optic sensory structures due to natural selection

North American Neo-Lamarckism

The first North American school of evolutionary theory, American neo-Lamarckism (Bowler 1983, Richardson and Kane 1988), emerged in the 1860s and persisted until the turn of the century. A diverse group of North American evolutionary biologists led by Alpheus Hyatt and Edward Drinker Cope provided a vigorous alternative to Darwinism. Both Hyatt and Cope originally studied groups of organisms that apparently showed a history of reduction and simplification rather than of increased complexity. Hyatt (1866) extensively studied fossil cephalopods, especially ammonites and nautiloids. These groups apparently shared a phyletic pattern of diversification and an increase in complexity followed by a loss of diversity and a decrease in complexity. Cope began with a different group of organisms that likewise indicated a trend for simplification over time—cave organisms (Cope 1872). It was the apparent "senescence" of these phyletic lines that required explanation by evolutionary biologists, Lamarckian and Darwinian.

For Lamarckians, species and evolutionary lineages that showed a decline in complexity posed a special problem. While we associate the terms *use* and *disuse* with Lamarckism, Lamarck himself regarded orthogenesis—change determined by factors internal to the organism, not by the environment—as the primary mechanism of evolution. He held that there was an innate drive toward perfection in a lineage, a concept that encompasses the idea of progress. Cope (1868) and Hyatt (1866) began with an orthogenetic view of evolution, assigning a secondary role to the effects of use and disuse and an even more subsidiary role to natural selection. As Richardson and Kane (1988) point out, orthogenetic evolution should lead to increased complexity. Cases of apparent simplification thus caused problems for those who argued from an orthogenetic viewpoint.

The key to the beginnings of American neo-Lamarckism was its commitment to Haeckel's biogenetic law: ontogeny recapitulates phylogeny (Haeckel 1874). For neo-Lamarckians the key aspect of evolution was not natural selection acting on existing variation, as Darwin supposed, but rather the origin of new characters. Genera were distinguished as having unique characters, and Cope (1868) pointed out his disagreement with Darwin in a paper cleverly entitled "On the Origin of Genera." As Stephen Jay Gould pointed out in *On-*

togeny and Phylogeny (1977), recapitulation requires terminal addition (addition of new features to ontogenetic stages in later stages of development) and acceleration (shortening of ontogeny in order to press newly acquired characteristics into earlier periods of development). If the origin of species (and especially genera, for neo-Lamarckians) can be explained by terminal addition and acceleration, then there must be a parallelism between allied forms, as well as some level of resemblance between the juvenile forms of more advanced genera and the adult forms of less advanced genera. Hyatt and Cope viewed this parallelism as the product of nonadaptive mechanisms controlling growth and development; the problem was to explain a decline in complexity—the simplification and apparent retardation of ammonite and nautiloid lines, and eye and pigment loss in cave animals—within this framework.

Cope put forward the "law of acceleration and retardation" as an alternative to natural selection:

> There are, it appears to us, two laws of means and modes of development. I. The law of acceleration and retardation. II. The law of natural selection . . . [While] natural selection operates by the "preservation of the fittest," retardation and acceleration act without any reference to "fitness" at all; . . . instead of being controlled by fitness, it is the controller of fitness. Perhaps all the characteristics supposed to generalized groups from genera up . . . have evolved under the first mode, combined with some interventions of the second, and . . . specific characters or species have been evolved by a combination of a lesser degree of the first with a greater degree of the second mode. (Cope 1868, p. 244)

The insistence by Cope (1872) and later by A. S. Packard (1888) that many cave species belonged to distinct genera is a reflection of the neo-Lamarckian view that real evolutionary change, that is, the origin of new characters, is in the origin of genera, not the origin of species.

With terminal addition, new characters are added toward the end of development. Acceleration brings them back to a time prior to reproduction, when they can be transmitted to offspring. Cope called this the "expression point." When this point is reached a new generic type is established. Retardation delays the expression of characters, which may eventually be pressed past the expression point. They are then not inherited.

Cope (1872) explained eye loss in cave animals in terms of retardation of growth. He thought that the attenuated growth rate resulting from retardation would at least diminish the size of eyes and eliminate other characters. The final elimination of eyes, Cope held, was due to disuse. Hyatt (1866) held many of the same views as Cope but differed on the cause of retardation. Unlike Cope, Hyatt thought both evolutionary advance and evolutionary decline were the result of acceleration, together with an internal, preprogrammed source of modification. In Hyatt's view, "senile" (retarded) features were analogs of juvenile features. A lineage begins as a juvenile form and adds characters with maturity. Finally, senile features are added and subsequently accelerated, as are all terminal additions. Eventually, all intermediate stages are crowded out, and the result is virtually indistinguishable from a juvenile form.

The distinguishing feature of Cope's and Hyatt's views in the 1860s and early 1870s was their commitment to orthogenesis, the orderly, linear development over time. Use and disuse played a secondary role in their work, although not as minor a role as natural selection. The ultimate cause of orthogenesis, at least in Cope's view (see Richardson and Kane 1988), was not mechanical but divine: "Our present knowledge will only permit us to suppose that the resulting and now existing kingdoms and classes of animals and plants were conceived by the Creator according to a plan of his own, according to his pleasure" (Cope 1868, p. 269).

Although strictly orthogenetic views such as Cope's and Hyatt's largely disappeared by the turn of the century, they are echoed in some later work of European cave biologists, particularly Albert Vandel (1964). Vandel held to an orthogenetic view of evolution in caves, denying any significant role for adaptation and arguing that cave animals were not blind because they lived in caves—they lived in caves because they were blind. Both Cope and Hyatt moved to the more classical neo-Lamarckian view, in which use and disuse were the motive force of evolution, and abandoned the theistic component for at least the origin of new characteristics.

Cope's early views on the evolution of cave organisms can be summarized in a manner parallel to Darwin's scheme:

1. IMMIGRATION: migration of surface forms into the cave habitat
2. DEGENERATION: atrophy and underdevelopment of visual organs because of retardation in development
3. INHERITANCE: transmission of orthogenetic (programmed) patterns to offspring
4. COMPENSATION: hypertrophy of extra-optic sensory structures due to acceleration of development

Ironically, Darwin relied more on disuse in the evolution of morphology in cave organisms than did Cope, at least in his early writings. It is the theism and orthogenetic views in Cope's work of the 1870s that both distinguish the rising school of North American neo-Lamarckism and make it seem strange to the modern reader.

This is not to deny the importance of the timing of developmental events in the ontogeny and phylogeny of cave animals. Retardation of development may well be important as a proximate but not an ultimate cause of some morphological characteristics of cave animals, especially neoteny in plethodontid salamanders (Bruce 1979), and of the loss of morphological complexity in general (Magniez 1985).

Alpheus Packard

It may be that, as Kane and Richardson (1985) claim, Cope was only a dabbler in biospeleology, and that only his early evolutionary work concerned cave organisms. Cope did publish four papers dealing largely or exclusively with cave animals and described a handful of cave species, mostly crustaceans, but then he returned to a study of fossils.

Alpheus Packard, on the other hand, concentrated on cave faunas throughout his scientific career, during which he published numerous species descriptions, over fifteen papers, and a book-length monograph (1888). Packard took advantage of the heightened interest in caves and cave exploration in North America (see Hovey 1882) to make extensive studies of caves and cave faunas. In 1888 Packard estimated that there were 500 caves in the U.S. (there are more than 40,000 now known! [Holsinger 1988]), and he personally visited at least several dozen of these. More important, he made cave faunas the centerpiece of his contributions to the neo-Lamarckian school of evolutionary biology:

> The study of the conditions of existence in caves is of special value, because such conditions are so unusual and the results upon certain organs so easily appreciated. It is by a study of life under unusual conditions that the attention is aroused and interest excited, and after acquiring experience in dealing with the more palpable, because somewhat abnormal, circumstances under which organisms exist, we can then more easily observe the effects of changes of ordinary conditions upon the organism. (Packard 1888, p. 142)

Packard is clearly convinced of the importance of environment and the unusual nature of the cave environment in particular for the study of evolution. It is not surprising that Packard emphasizes the role of use and disuse rather than the much more idealized role of orthogenesis.

Packard was convinced of the efficacy of use and disuse as a mechanism not only for degeneration but for elaboration as well. He argued, as did Darwin, that eye loss was due to disuse, but he disagreed with Darwin on the causes of the elaboration of extra-optic sensory structures: "We may, with Darwin, for convenience, use the phrase 'natural selection' to express the process by which cave fauna was produced, but such a term to our mind expresses rather the result of a series of causes than a vera causa itself" (Packard 1888, p. 138).

Packard held that the true causes were environmental changes that accompanied isolation in caves. These changes, resulting in new patterns of use and disuse of organs and followed by the inheritance of acquired characters, were the "efficient causes" of the evolution of morphology of cave animals. He claimed that there was no "struggle for existence." Packard's views can be summarized in the following four steps:

1. IMMIGRATION: migration of surface forms into the cave habitat
2. DEGENERATION: atrophy and underdevelopment of visual organs because of lack of use
3. COMPENSATION: hypertrophy of extra-optic sensory structures due to use
4. INHERITANCE: transmission of environmentally induced reductions and elaborations to offspring

Unlike either Darwin or Cope in his Lamarckian phase (see above), Packard and other neo-Lamarckians held that the degeneration of some structures and the hypertrophy of other structures were due to the same cause—namely, use and disuse followed by hereditary transmission of environmentally induced changes. More than anyone before him, Packard stressed not only structural reductions but structural elaborations as well. He held that eyelessness *per se* was sufficient neither to define the morphological syndrome of cave animals (troglomorphy) nor to define the genera erected for cave animals. In a defense of the retention of the genus *Caecidotea* for cave species of *Asellus* (originally thought to be related to *Idotea* rather than *Asellus*), he states: "It should be observed that not only are *Caecidotea stygia* and *Caecidotea nickajackensis* without eyes, but that the body and appendages also differ a good deal from any of the known species of *Asellus*" (Packard 1888, p. 30). He devoted considerable attention to the anatomical changes accompanying both eye degeneration and the hypertrophy of extra-optic sensory structures.

In practice, Packard's main dispute with Darwinians revolved around questions of time. Packard held that the morphological changes observed in cave animals, while extensive, occurred rather quickly. More than anything else, it was the apparent slowness of natural selection, the numberless generations in Darwin's phrase, that Packard perceived as wrong. To Packard the action of use and disuse was not only the "vera causa," it worked rapidly. He held that after only a few generations, perhaps less than ten, the compound eye of isopods would be reduced to a rudiment (ibid., p. 118). He also held that the degeneration of the eye either happened faster than or before the elaboration of extra-optic sensory structures. For example, he discusses cave-dwelling species of the amphipod genus *Crangonyx* that show little modification except for eye reduction.

Although Packard does not explicitly state that the amount of eye degeneration can be used as an evolutionary clock, albeit a rather fast one, this was clearly his intention. He discusses at length the evidence (largely illusory) that the fauna of what he held to be a very recently formed cave in Utah had slightly reduced eyes. Likewise, he discusses apparent exceptions to his claim that life in darkness must result in eye degeneration. For example, he pointed out that various camel crickets found in caves retain eyes because they periodically leave the cave to feed. At the end of the nineteenth century, the general scientific view held that the earth and its creatures were much younger than we believe them to be. Thus Packard believed that the Mammoth Cave, whose age is currently estimated to be 2 million years (Palmer 1989), was no more than 7,000 to 10,000 years old.

Although neo-Lamarckism fell into disrepute around the turn of the century, the idea that the amount of eye degeneration was a measure of evolutionary time did not. The pioneering studies of T. L. Poulson (1963) on adaptation in cave fish explicitly used the amount of regressive evolution as a relative measure of how long a species had been isolated in caves. In this case the degeneration was not claimed to be the result of disuse, of course, but rather the result of the accumulation of structurally reducing, selectively neutral mutations. This idea of a morphological clock resulting from an underlying mutational clock has been explicitly employed by Wilkens in his studies of cave fish (Wilkens 1973, 1986).

Lamarckism in Disarray

With the turn of the century came a wholesale reformulation of the problem of speciation, and the neo-Lamarckian tradition fell into disarray. On the one hand, the rediscovery of Mendel's results combined with Weismann's insistence on the independence of germ cells and somatic cells made the inheritance of acquired characters unnecessary and untenable. In addition, the discovery of radioactive decay had undermined the general challenge to Darwinism by greatly extending Kelvin's estimates of the age of the earth. All arguments against Darwinism based on the claim that there had been insufficient time for natural selection to operate were rejected.

It was in this atmosphere than A. M. Banta (1907) speculated on the evolution of cave animals. The problem for Banta was how to explain the evolution of troglomorphy within the framework of natural selection, despite the lack of any clear adaptive function of regressed characters. Banta saw three basic questions pertinent to understanding the evolution of cave animals: How did animals come to be in caves? What were their morphological characteristics when they first immigrated into caves? How have they arrived at their present condition?

Banta rejected Packard's claim of the rapid evolution of cave-associated traits. He suggested instead that the evolution of cave animals was much more gradual, starting with ancestors that frequented caves through "voluntary immigration" rather than accidental isolation. This argument anticipates the current argument (see Rouch and Danielopol 1987) about whether animals become isolated in caves as a result of active invasion by relatively large populations or as a result of the "stranding" of a small number of individuals. Indeed, we return to this question in a later chapter.

Banta further postulated that many of these ancestors may have been predisposed to cave life if they favored dark or shady habitats on the surface. This explanation, in Banta's view, provided two potential avenues for cave adaptation. Some "voluntary migrants" may begin by living in cave entrances and gradually, in succeeding generations, modify further until they can survive in the deepest recesses. In addition, he suggested that some other species may have already become so highly modified for cave life that they are initially able to survive in the deep cave, even though they are surface-dwellers.

Superficially, Banta's scenario may seem to resemble the modern view that successful cave immigrants are often preadapted or exapted for cave life. Closer scrutiny suggests that his view was orthogenetic:

> Animals do not possess degenerate eyes and lack pigment because they are cave animals. The eyes have in many cases degenerated and the color disappeared before they entered caves. They are cave animals because their eyes are degenerate and because they lack pigment . . . They are isolated in caves and other subterranean abodes because they are unfit for a terranean life and caves are among the possible habitats. (Banta 1907, p. 99)

Thus, Banta denied that degenerate characters were adaptations to the cave environment, maintaining instead that they were maladaptations to surface life.

There is some resemblance between Banta's view of maladapatation and the orthogenetic view of racial senescence held earlier by Hyatt. Banta, however, was aware that the theory of inheritance of acquired characters had been undermined. Ultimately, in the case of eye and pigment loss, Banta could only conclude that these losses appear to be consistent with "the theory of the cumulative effect of determinate variations" (Banta 1907, p. 104), but he could provide no mechanism.

The theory that some organisms may have predisposed to a cave environment prior to entering caves also led Banta to be skeptical of using the extent of modification in cave-related characters as an index of the age of cave fauna. Banta believed that equating the degree of regressive evolution with the time spent in caves tended to support the now untenable theory of disuse.

Having dispensed with disuse but returning to orthogenesis as an explanation for reduced characters, Banta arrived, in the concluding section of his monograph, at an argument that presages modern views about the evolution of elaborated characters in cave environments. He suggested that attenuated appendages and increased number and development of sensory papillae and setae "may be due to individual adaptation or . . . to natural selection tending to eliminate all individuals which do not possess variations of assistance to the animal under its unusual conditions" (ibid., p. 104).

Banta represents a transition from a neo-Lamarckian view to a Darwinian view, but his Darwinism has Lamarckian components as well. He clearly ascribes a greater role to natural selection and a diminished role to disuse than Packard envisioned, and, unlike Packard, he saw no connection, even in terms of a common cause, between regressed and elaborated characters.

Neo-Darwinism

Neo-Darwinians, particularly Christiansen (1961, 1965) and Poulson (1963), have studied extensively the second characteristic of cave animals, hypertrophy of extra-optic sensory structures and other increases in morphological complexity. In a series of studies of cave Collembola, Christiansen convincingly demonstrated a striking morphological parallelism and convergence in the claw structure of cave species; unlike surface-dwelling species, collembolans that inhabit caves have claws that allow them to move effectively across water surfaces and wet mud, two common cave microhabitats. These morphological changes are summarized in Figure 2.2. Poulson made similar

Fig. 2.2. Morphological and positional changes in the claws of three collembolans (family Entomobryidae): a highly specialized, cave-limited *Pseudosinella (right);* a *Pseudosinella* species found both in and out of caves *(middle);* and a primarily surface-dwelling species *(left).* Individuals with claws like those on the right are able to move easily across wet mud and water, as shown here; the others are not. Modified from Christiansen (1965).

studies of cave fish, and he too argued that natural selection was a major factor in molding the morphology of cave animals. Neither addressed directly the causes of eye and pigment loss, but they both used the degree of degeneration as a crude estimate of the length of time of isolation in caves.

Impressive though the work of Christiansen and Poulson was, it was outside the main thrust of neo-Darwinian interest in cave animals. If eye degeneration in cave animals is a Lamarckian dream, the geographical distribution of cave animals would seem to be, if not a neo-Darwinian ideal, at least a "Mayrian" dream of allopatric speciation. The typical absence of closely related geographically proximate ancestors and the restricted ranges of most cave species led to a classic paradigm of invasion and subsequent speciation among the isolated cave populations (Barr 1968, Juberthie 1989). Much attention has been focused on the details of speciation of cave organisms, including active versus passive movement into caves (Rouch and Danielopol 1987), vicariance versus underground dispersal (Kane, Culver and Jones 1992), and non-allopatric speciation (Wilkens and Huppop 1986).

Among general evolutionary theorists, the loss of eyes of cave animals and their genetic differentiation have provoked the most recent interest. For example, Prout (1964) and Chakraborty and Nei (1974) have developed population-genetic models to explain eye loss. Molecular techniques have been applied to questions of genetic differentiation of cave animals (Caccone and Powell 1987).

As biologists have searched for greater generality, much of modern cave biology has likewise focused in one way or another on the questions of eye loss. The possible consequences of neutral mutation on eye reduction have been extensively studied by Wilkens and his colleagues (Wilkens 1988) but without a similar study of other aspects of morphology. Many European biologists, following issues first raised by Racovitza (1907) and more forcefully taken up by Motas (1958), have looked to all subsurface communities where eye loss is common. The idea of the underground ecosystem (Rouch 1977), with consideration given to animals living in large cavities (for example, caves) and small cavities (for example, underflow of streams), has considerable appeal. Communities of this type share a dependence on external food sources and an absence of sunlight. On the ecological level, they are often united in the same underground drainage basin and share a resource base (Rouch 1986), but on the evolutionary level, it is not at all apparent that there is an underlying unity among the organisms in the communities. Beyond a general eyelessness, the morphology of animals living in large subterranean cavities and small subterranean cavities is not that similar. For example, the two isopods shown in Figure 2.3 are typical of the two habitats. The one from the underflow of streams (a hyporheic habitat) is fusiform and generally small. The cave-dwelling isopod, in contrast, is large and has long appendages. Among Collembola, eyeless species of *Onychiurus* inhabit deep soil, whereas eyeless *Pseudonsinella* inhabit caves. Yet the two genera (Fig. 2.4) represent morphological extremes among the Collembola; the only feature they share is eyelessness.

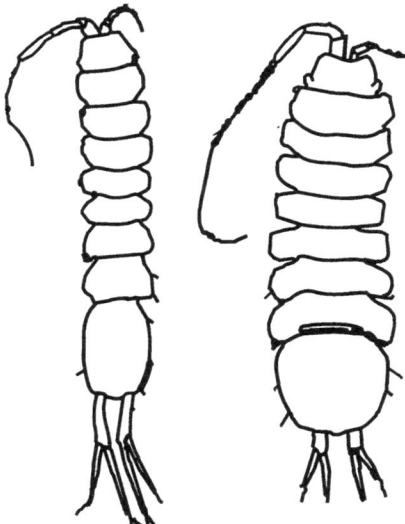

Fig. 2.3. Proasellus albigensis, an isopod from hyporheic habitats *(left)*, is 5 mm long; *Proasellus vandeli*, an isopod from caves *(right)*, is 4 mm long. Modified from Henry and Magniez (1983).

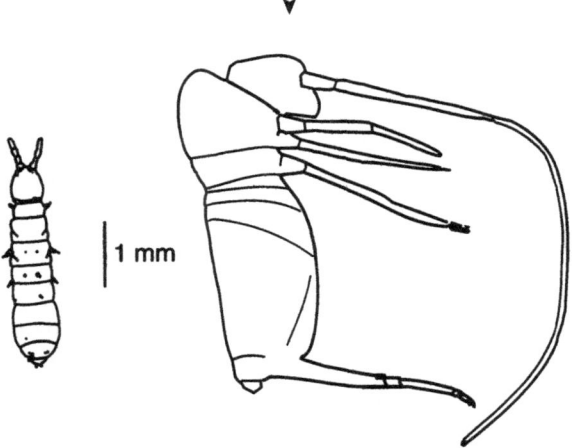

Fig. 2.4. A typical soil collembolan *(Onychiurus armatus, left)* and a highly troglomorphic collembolan *(Pseudosinella christianseni, right)*. Drawings courtesy of K. A. Christiansen.

Interesting as this work is, it misses what to us is the central question of cave biology and the most interesting question in evolutionary biology for which cave organisms may serve as case studies. This is, what factors, both proximate and ultimate, have molded the morphology of thousands of species of cave organisms in a similar way? Cave organisms are not primarily of interest because they are eyeless. Cave organisms are not primarily of interest because they are bizarre. They are primarily of interest because their evolutionary history—a history of similar selective pressures produced by similar and unusual environmental conditions, and similar potentials for genetic differentiation resulting from similar levels of habitat separation and distinctness—offers us the possibility of making predictions about genetic and morphological patterns of evolutionary change.

The Evolution of Troglomorphy

E. G. Racovitza, in a seminal work (1907) that has a continuing impact on studies in cave biology, especially in Europe, listed the following as characteristics of "ideal" cave animals:

1. Lack of pigment, even in the presence of light
2. Eyelessness, or having only rudiments of the optic apparatus
3. Hypertrophy of other sense organs, especially tactile organs, as compensation for blindness; hypersensitivity to environmental vibrations
4. Fragility of organs
5. Elongation and flattening of body
6. Photophobia; extreme sensitivity to temperature variation and to desiccation
7. Absence of regular cycles of activity and reproduction

Modern biologists would certainly quarrel with some of these characteristics, especially the seventh, but the major point is that cave organisms share a set of common morphological and behavioral features well beyond simple eye and pigment loss. Christiansen (1962) coined the very useful term *troglomorphy* to describe the morphological syndrome associated with life in caves:

> Troglomorphs are forms where the body is clearly modified for cave existence and which are quite different from all normal non-cavernicolous animals. Such forms are, on the whole, easily identifiable as cave animals and have many common traits even when they belong to different taxonomic groups. The members of this group show striking convergent evolution and eventually evolve forms extremely different from their ancestral forms living outside caves. (Translated by K. Christiansen)

Christiansen (1992) has revisited the question of the convergent characters of cave faunas and lists an extensive set of morphological, physiological, and behavioral characters shared by cave organisms (Table 2.1). For Christiansen and for us, the idea of convergent and parallel evolution of cave faunas is pivotal in defining troglomorphy.

It should be noted that troglomorphy is not universal among subterranean organisms. As noted before, organisms living in small subterranean cavities show very different patterns (see Figs. 2.3 and 2.4). Some cave organisms do not conform to the standard description because they live in energy-rich environments such as guano piles and

Table 2.1. Major characteristics of cave organisms relative to surface-dwelling organisms. Slightly modified from Christiansen (1992).

Morphological
  Specialization of sensory organs (chemoreceptor, hygroreceptor, thermoreceptor, pressure receptor)
  Elongation of appendages
  Reduction of eyes, pigments, and wings
  Cuticle thinning (in terrestrial arthropods)
  Foot modifications (in Collembola and planthoppers)
  Scale reduction or loss (in fish)

Ecological and physiological
  Slowing metabolism
  Starvation resistance
  Relaxation and degeneration of Circadian rhythms
  Lowered fecundity
  Increased egg volume
  Increased life span

Behavioral
  Decreased aggregation (in Collembola)
  Reduced reaction to alarm substance (in fish)
  Increased sensitivity to vibration
  Reduced intraspecific aggression

have not been in caves long enough for troglomorphic features to develop fully (Culver 1982).

For any group of related species or populations in any habitat, a list such as that in Table 2.1 can be compiled. Indeed, the elaborated characters, such as specialization of sensory organs, are apomorphic characters, required for cladistic analysis. Typically, shared apomorphic characters (synapomorphies) are used to indicate phylogenetic relationship. Independent evolution of apomorphic characters (homoplasy in part) is minimized in the construction of phylogenetic trees. The reduced characters, such as pigment (Table 2.1), are character reversals. In the construction of trees, the optimal tree is determined, in part, by minimizing these character reversals as well as by minimizing independent evolution of apomorphic characters. What makes the evolution of cave organisms of general interest is the claim of

Christiansen, Poulson, and other neo-Darwinians that the hallmark of cave organisms is the convergent and parallel evolution of both elaborated characters and reduced characters. If the absence of natural selection is, as Lewontin said, a taxonomist's dream, then cave organisms are a nightmare. The morphology of cave organisms is characterized by homoplasy and character reversal, the very features cladists attempt to minimize.

We can now restate the central question of interest—what are the proximate and ultimate factors responsible for the evolution of troglomorphy (homoplasy)?

Modern Studies of Evolution in Caves

Most discussion of the evolution of the morphology of cave organisms has focused on the development of reduced characters, which is generally termed "regressive evolution." Hypotheses to explain regressive evolution in cave organisms are myriad and largely nonselectionist. In a recent symposium devoted to the subject of regressive evolution (Culver 1985), direct selection, pleiotropy, neutral mutation, allometric effects, and differential migration into caves were all invoked as potentially major factors in regressive evolution. Barr (1968) and Culver (1982) give extensive reviews of the subject.

We take a different approach. The very separation of reduced characters from elaborated characters in the study of the morphology of cave organisms is misleading; it leads to a set of special hypotheses about a subset of the question. To reiterate, the question is how *both* elaborated and reduced characters (the troglomorphic syndrome) evolved, not just how reduced characters evolved. An emphasis on reduced characters reinforces what we believe is an incorrect notion that cave organisms are special, different from other organisms. What is special about cave organisms is a difference in degree, not a difference in kind: the apparently high degree of parallelism and convergence of both elaborated and reduced characters.

Our approach is to examine the role of natural selection in molding the biology of cave organisms—that is, the role and importance of adaptation. Throughout the history of the study of evolution, the twin concepts of adaptation and natural selection have been the focus of often intense debate. The great debates of evolutionary biology have involved at their core a debate about natural selection. The debate between LaMotte (1951) and Cain and Sheppard (1954) about color banding in *Cepaea nemoralis* and the debate over the causes of allozyme variation (Lewontin 1972, Gillespie 1991) are but two prominent examples. In fact, the debate over the role of natural selection in the evolution of cave organisms is over the adaptive significance of troglomorphy. Thus, the adaptationist studies of Christiansen (1961, 1965) and Poulson (1963) were implicit attacks on the orthogenetic views of Thines (1969) and Vandel (1964). Vandel made a most explicit attack on adaptation: "The notion of adaptation has become such an obsessive issue that it could have been written that depigmentation and eyelessness represent adaptations to cave life. You might as well say that congestion, rheumatism and far-sightedness are adaptation to old age" (Vandel 1964, p. 563, translated by the authors).

Over the past decade there has been renewed interest in adaptation and natural selection from a variety of perspectives. Population geneticists have developed new ways of testing for natural selection (see Lande and Arnold 1983, Endler 1986, Schluter 1988); molecular geneticists are developing new techniques for determining the genetic structure of populations (summarized by Hillis and Moritz 1990); evolutionary biologists have developed techniques for removing phylogenetic effects from data (see, for example, Lynch 1991); and philosophers of science have provided fresh insights into these issues (see especially Sober 1984, Levins and Lewontin 1985, and Brandon 1990).

What are the hypotheses that can account for the following idealized case? A population of cave arthropods differs from a closely related surface population in having no eyes and elongated antennae. We begin by considering causes of elongated antennae. Obviously, this feature may be the result of selection for elongated antennae (and increased extra-optic sensory structures in general) in the cave environment, resulting in adaptation. Second, it may be the result of selection, but selection for elongated antennae for some other function (exaptation), selection for elongated antennae in some other environment (preadaptation), or a correlated response to selection for some other character such as overall size. Third, elongated antennae may be the result of neutral mutation and genetic drift. Lande (1976) and Lynch (1990) have set forward the expected rates of morphological change under neutral mutation.

We can now see one of the advantages cave organisms have as subjects for the study of natural selection and adaptation: the idealized pattern of small-eyed, large-antennaed descendants in a distinct and separate environment (caves) is repeated over and over. Natural selection can thus be analyzed using two powerful sets of analytical tools. The first set is that of population genetics, including traditional quantitative genetics (for example, heritability), direct measurement of selection intensity, and estimation of genetic distance and relatedness using molecular-genetic data. The second set is that of systematics and cladistics, including information on morphometrics and homoplasy. The study of homoplasy (false homology [Wake 1991]) is a particularly powerful method of detecting natural selection. This analysis is summarized in Figure 2.5.

When we turn to causes of eye reduction, we see the second major advantage of cave organisms in the study of natural selection and adaptation: reduced eyes and other morphological and behavioral reductions may represent a tradeoff that makes possible the elaboration of extra-optic sensory structures. The idea of ecological and evolutionary tradeoffs permeates the literature, but it is perhaps most thoroughly developed in ecological contexts such as life-history evolution (Stearns 1980). When it is applied to a cave population, the question is whether, at some level of organization, the degeneration of eyes is required for the evolution of elaborated extra-optic sensory structures, such as antennae. This evolutionary tradeoff may be the result of energy economy (Poulson 1963), competition for neurological connections to the brain (Jones and Culver 1989), or molecular-genetic "noise suppression" (Regal 1977). If evolutionary tradeoffs

28   Caves as Evolutionary Laboratories

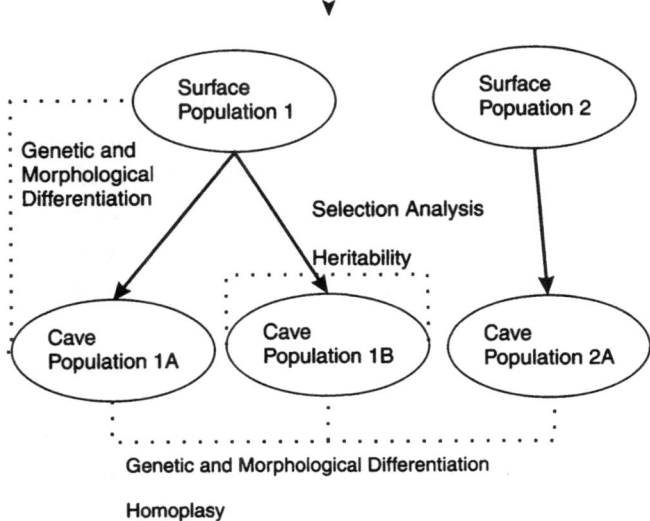

Fig. 2.5. Summary of various analytical possibilities for the study of evolution of cave animals. Each population can be analyzed separately (for evidence of selection and heritability), in pairs of cave and surface populations (differentiation), and in groups of cave populations (differentiation and homoplasy).

are important, then they are most likely important in cave organisms, where the likelihood of tradeoffs is high. It is high not only because of the presence of repeated morphological patterns, but because of the environmental extremes of caves.

There are, of course, other explanations for eye loss. All the reasons given for elaboration of a structure may hold for the loss of a structure. The hypothesis that a loss may be due to neutral mutation has a long history among cave biologists (Kosswig and Kosswig 1940, Culver 1982, Wilkens 1988); neutral mutation has sometimes been considered by its proponents as a major evolutionary force in cave organisms extending well beyond reduced morphology to the apparent loss of many behaviors as well (Parzefall 1992). A third hypothesis, quite specific to caves but nonetheless with some credence, is that differential migration of forms with reduced eyes into caves and migration of eyed forms out of caves results in differentiation and eye loss in cave populations (Ludwig 1942, Fong and Culver 1985). These hypotheses are summarized in Table 2.2.

Table 2.2. Hypothesized causes of morphological change in cave animals.

Elaborated structures	Reduced structures
Adaptation	Adaptation (evolutionary tradeoff)
"Other selection": exaptation pre-adaptation correlated response	"Other selection": exaptation pre-adaptation correlated response
	Neutral mutation
	Differential migration

Summary

Most cave organisms share certain characteristics: a bizarre morphology, including reduced eyes or an absence of eyes; a highly restricted geographic range with no closely related surface-dwelling relatives; and extreme longevity and low reproductive rates. To Lamarck, cave organisms were just a minor example of the consequences of disuse. Darwin also focused on the blindness of cave animals, which he attributed to disuse of the optic system. For other organisms, however, Darwin did argue that losses could be the result of evolutionary tradeoffs.

North American neo-Lamarckians, especially A. S. Packard, attempted to explain all the morphological changes cave animals display—the eye and pigment reductions and the elaboration of extraoptic sensory structures. Packard held that both elaborations and reductions resulted from the effects of use and disuse and the hereditary transmission of these changes.

Neo-Darwinian cave biologists, especially Christiansen and Poulson, argued that natural selection was the essential mechanism for both the reduction of some characters and the elaboration of others. Christiansen termed the suite of parallel and convergent morphological changes observed in cave animals "troglomorphy." Other evolutionary biologists have focused on explanations for eye loss that do not involve natural selection, especially neutral mutation.

Cave organisms offer two major advantages in the study of adaptation—the highly repeated nature of the morphological changes in many taxonomic groups in many areas and the apparent evolutionary tradeoffs resulting from the unique features of the cave environment, especially the absence of light.

Selected References

Barr, T. C. 1968. Cave ecology and the evolution of troglobites. *Evolutionary Biology* 2:35–102. An extensive review of theories of evolution of cave animals.

Botosaneanu, L. [ed.] 1986. *Stygofauna mundi.* Leiden: E. J. Brill. This "bestiary" of the subsurface provides a thorough introduction to aquatic cave and subterranean fauna in general.

Christiansen, K. A. 1965. Behaviour and form in the evolution of cave Collembola. *Evolution* 19:529–532. A classic case study of adaptation of cave organisms.

Culver, D. C. 1982. *Cave life: Evolution and ecology.* Cambridge, Mass.: Harvard University Press. An extensive discussion of the selectionist-neutralist controversy concerning the evolution of cave animals.

——— ed. 1985. Regressive evolution. *National Speleological Society Bulletin* 47, No. 2. Proceedings of a symposium, with a wide range of views expressed, devoted to regressive evolution of cave animals.

Packard, A. S. 1888. The cave fauna of North America, with remarks on the anatomy of brain and the origin of the blind species. *Memoirs of the National Academy of Sciences (USA)* 4:1–156. Packard's monograph not only provides a detailed statement of the neo-Lamarckian view of evolution of cave animals, it also offers a fascinating glimpse at nineteenth-century cave biology in general.

Racovitza, E. G. 1907. Essai sur les problèmes biospéologiques. *Archives de Zoologie Expérimentale et Générale* 4:371–488. An influential essay that reads like a research program for the study of subterranean life.

Vandel, A. 1964. *Biospéologie. La biologie des animaux cavernicoles*. Paris: Gauthier Villars. Aside from his highly idiosyncratic view on evolution, Vandel provides a good survey of both aquatic and terrestrial cave fauna.

Wilkens, H. 1988. Evolution and genetics of epigean and cave *Astyanax fasciatus* (Characidae, Pisces). Support for the neutral mutation theory. *Evolutionary Biology* 23:271–367. An extensive argument in support of the importance of neutral mutation in the evolution of cave animals.

3 *Gammarus minus* as a Model Organism

The troglomorphic syndrome of cave organisms is an enticing subject for the study of natural selection and adaptation. Ironically, most troglomorphic cave organisms are the least suitable for detailed evolutionary studies. They often lack variation in troglomorphic characters, and the lack of close surface relatives for phylogenetic comparison makes it difficult to determine the ancestral states of troglomorphic features. Here we introduce an exception: *Gammarus minus*. A freshwater amphipod crustacean, *G. minus* has several advantageous features that render it particularly amenable for the study of the evolution of troglomorphy.

Obstacles to the Study of Cave Organisms

In spite of their obvious interest from an evolutionary point of view, cave organisms have played a relatively minor role in modern evolutionary biology. Indeed, except for their study by North American neo-Lamarckians, cave organisms have had little impact in evolutionary biology in general. Why hasn't the study of cave organisms lived up to the potential noted in Chapter 2?

First and foremost, despite claims to the contrary (see Barr 1966), cave biology remains dominated by taxonomic studies. The discovery of new subterranean faunas from a variety of geographic areas, such as Mexico, Thailand, and Hawaii to name only a few, and the discovery of new subterranean faunas from a variety of habitats other than caves have continued to occupy the attention of many able cave biologists.

Second, because of the taxonomic distribution of cave organisms, phylogenetic analysis of the evolution of cave faunas is difficult. On the one hand, there are many groups of cave organisms that are in exclusively subterranean genera or families. Examples from the North American cave fauna, each with more than 50 described species, include the amphipod genus *Stygobromus* and the beetle genus *Pseudanophthalmus*. Cladistic analysis has largely focused on the level of groups that are entirely subterranean (see Notenboom 1988, Boutin, Messouli, and Coineau 1993). On the other hand, some groups have invaded caves only once (for example, *Proteus*), and it is difficult to argue beyond the historical uniqueness of this event.

Third, many troglomorphic features are quantitative rather than qualitative. Extra-optic sensory structures increase in size, eyes decrease in size, and appendages elongate. Qualitative changes that do occur are often losses (pigment, eyes, etc.) rather than gains. Thus, the study of troglomorphy must involve the techniques of morphometrics, techniques not familiar to most taxonomists. On the other hand, morphometricians have shown little interest in troglomorphy. The quantitative changes associated with troglomorphy are, at the present level of our understanding, not changes in well-integrated features like the vertebrate skull, a topic of immense interest to morphometricians. Understanding complex, well-integrated features requires the development of new techniques such as landmark analysis (Rohlf and Bookstein 1990). Analysis of troglomorphy has for the most part utilized more traditional techniques (see Christiansen and Culver 1968, Culver 1987), to the extent that morphometrics has been used at all.

Fourth, because of their long life cycles and low reproductive rates (Culver 1982), most cave animals have not proved amenable to laboratory study, such as population cage experiments or artificial selection experiments. Indeed, the only experiment of which we are aware that attempts to select against eyes was an experiment by Payne (1911) exposing *Drosophila melanogaster* to 69 generations of continuous darkness.

Finally, many highly troglomorphic cave animals are rare, morphologically uniform, and in some cases genetically uniform (Culver 1982). They seem to represent little more than evolutionary cul-de-sacs, arrived at after long periods of isolation in caves. Evolutionary tradeoffs have long since been resolved, and all that remains is the end result, with no information on the process.

Variable Cave Organisms as Model Systems

Not all cave animals are blind and eyeless, nor are they all morphologically and genetically homogeneous. Troglomorphic characters are not limited to species isolated in caves for millions of years with no close surface relatives. Among cave biologists, the Kosswigs were perhaps the first to turn from exotic and bizarrely specialized evolutionary endpoints to a much more mundane cave animal—populations of the isopod *Asellus aquaticus* (Kosswig and Kosswig 1940). It was the Kosswigs and another German biologist (see Ludwig 1942) who focused attention for the first time on the population genetics of cave animals. They recognized two important advantages of an animal like *Asellus aquaticus*. First, it displays variation in troglomorphic characters. This variation is critical for any experimental analysis of adaptation and of invasion of caves. Ludwig used *Asellus aquaticus* to investigate the possibility that the early stages of differentiation of cave and surface populations resulted from the differential migration of reduced-eyed forms into caves (Janzer and Ludwig 1952), a hypothesis that obviously requires variable populations. The fact that the hypothesis is wrong, at least for *Gammarus minus* (Fong and Culver 1985, Vawter, Fong and Culver 1987), is beside the point. Second, species like *Asellus aquaticus* invaded caves repeatedly. Kosswig (1965) took advantage of these replicated natural experiments to argue for the importance of selectively neutral mutations, especially those involving the eye, in the differentiation of cave organisms. As with Ludwig's theory, we have argued that Kosswig's view of evolution of cave animals is also mistaken (Jones and Culver 1989, Jones, Culver and Kane 1992). The collective importance of Ludwig's and

Kosswig's work is the emphasis on variation and on the need for testable hypotheses.

Perhaps the most-studied cave organism is the Mexican cave characin *Astyanax fasciatus* (formerly *Anophtichthys*). Widely used by neurobiologists and physiologists as a model blind animal because of its widespread availability (for example, Voneida and Fish 1984), it is also a well-studied case of regressive evolution (Wilkens 1988). Extensive regressive evolution occurred in *Astyanax* several times independently, when different surface stream populations were isolated in caves by stream piracy (Mitchell, Russell, and Elliott 1977). There has been considerable reduction of both eyes and pigment in these fish. In some caves, eye loss has proceeded to the point where there is no optic nerve and hardly anything left of the eye capsule. Accompanying the reduction in eyes and pigment are changes in the skull and pineal gland (Culver 1982). Compared with its reduced features, elaborated features of *Astyanax* are less dramatic and appear to be limited to taste organs (Wilkens 1988).

Gammarus minus shares with *Asellus aquaticus* and *Astyanax fasciatus* several critical features. First, presumed conspecific surface-dwelling populations are available for direct phylogenetic comparison. Second, repeated invasions into caves result in natural replication. Third, relatively large surface- and cave-dwelling populations make genetic analysis feasible. Fourth, there is considerable variation, within and among populations, in traits whose development is clearly correlated with the differences between cave and surface environments, that is, troglomorphic traits.

Basic Morphology of *Gammarus minus*

Gammarus minus was originally described in 1818 by the early American naturalist Thomas Say from specimens collected from a spring run near Lancaster, Pennsylvania. It is in many ways a typical *Gammarus*, the basic external morphology of which is shown in Figure 3.1. *G. minus* exhibits the appendage differentiation characteristic of all but the most primitive Crustacea. There are two pairs of an-

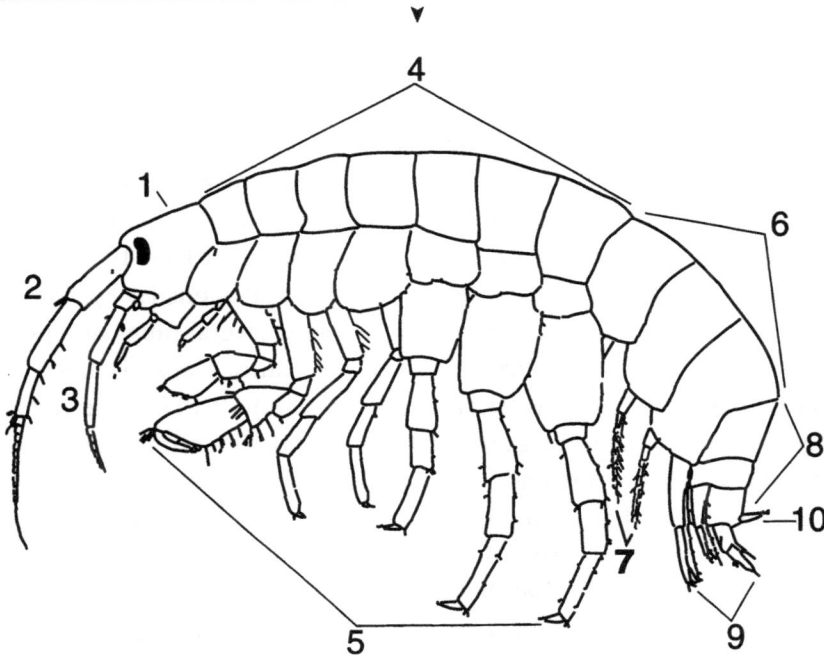

Fig. 3.1. The principal external features of *Gammarus:* *(1)* head; *(2)* antenna 1; *(3)* antenna 2; *(4)* pereonites 1–7; *(5)* pereopods 1–7 (including gnathopods 1 and 2); *(6)* pleonites 1–4; *(7)* pleopods 1–3; *(8)* uronites 1–3; *(9)* uropods 1–3; and *(10)* telson. Modified from Holsinger (1972).

tennae, mandibles, maxillae, seven pairs of trunk appendages (pereopods), the first two of which (gnathopods) are modified for grasping, three pairs of filamentous pleopods, and three pairs of uropods. The common name for amphipods, sideswimmers, is particularly apt for *Gammarus minus*. When not resting on their sides, individuals swim on their sides or back with rapid thrusts of the abdomen. As is typical of *Gammarus* in freshwater streams, *Gammarus minus* is a shredder. The trophic position of *G. minus* in cave communities, however, is often much more flexible (see Chapter 4).

The first cave population of *Gammarus minus* to be discussed was from Organ Cave in Greenbrier County, West Virginia. Shoemaker (1940) designated a new variety, *tenuipes*, for this population. He noted six distinguishing features (see Fig. 3.1 for terminology): (1)

greatly reduced eyes; (2) slender gnathopods and pereopods; (3) reduced number of spines on urosome; (4) lengthened inner ramus of third uropod; (5) rounded corners of lateral lobes of the head; and (6) a "weaker and more delicate appearance." Shoemaker makes no comment on why he designated this population a variety, especially given what to him appeared to be a very distinct population. We can only speculate that he did not describe it as a species because its restricted geographic range was well within the range of "normal" *G. minus*. He mentions that there were collections of two other populations of this variety from nearby caves.

Hubricht was probably the first amphipod biologist to see cave populations of *Gammarus minus* in the field. In addition to the characters listed by Shoemaker, Hubricht (1943) noted that *tenuipes* populations were bluish rather than brownish in color and that the antennae were long. Hubricht suggested that the population had immigrated into caves long ago and argued from microdistributional data that it was clearly distinct from spring populations of *G. minus*. He pointed out that *G. minus* populations rarely if ever extend from a cave stream out through a resurgence. He made no suggestions concerning the taxonomic status of *tenuipes* populations. Bousfield (1958) briefly considered *G. minus* and suggested that *tenuipes* populations be described as a separate species.

Holsinger and Culver (1970) took a closer look at cave populations and pointed out the high within- and among-population variability of *G. minus* populations. They re-examined the characters Shoemaker and Hubricht suggested as distinguishing *tenuipes* populations. Since Shoemaker looked at only one population in detail, it was impossible to determine which features were troglomorphic and which were simply a local population variant. Holsinger and Culver, from a study of *tenuipes* populations from two separate geographic areas, concluded that the following characters were troglomorphic:

1. Eye reduction
2. Lengthening of antennae
3. Increase in overall length

4. Increase in apparent fragility due to lengthening of pereopods and uropods
5. Shift in color from brown to blue and even white

They further concluded that since within-population variation was considerable and since almost every considerable intermediate existed somewhere in the range of the species, the variety *tenuipes* had no taxonomic or evolutionary validity. Better sense of the variation in *G. minus* can be made be placing the variation on a geographic template, which we do in Chapter 5.

Extent of Troglomorphy in Gammarus minus

The extremes in external eye morphology of *G. minus* are shown in Figure 3.2. The number of externally observable ommatidia shows a twenty-fold difference between the extremes: most spring populations average about 40 ommatidia per eye, and many cave populations have less than 2 ommatidia per eye (Holsinger and Culver 1970). The differences in internal structure of the eye between cave and spring populations are at least as great as the external difference. When viewed in cross section, the eye of highly modified cave populations appears disorganized; some components are missing and the remaining components jumbled (Fig. 3.3). The pattern of eye reduction extends to changes in the central nervous system. The optic lobe of individuals from Organ Cave is about half the size of the optic lobe of individuals from Organ Spring (Figs. 3.4 and 3.5).

These morphological changes are reflected in behavioral responses to light (Table 3.1). Whereas individuals from both spring and cave populations avoid light, *G. minus* from springs are much more strongly photophobic (Vawter, Fong, and Culver 1987). Spring populations seem to be more averse to light in part because cave populations do not have an integrated visual system with which to respond to light.

Cave populations also show increases in size and elaboration of some structures. The reduction in optic-lobe size in individuals from

Table 3.1. Reaction to light of different populations of *Gammarus minus* from Greenbrier County, West Virginia. Animals were placed in a tube darkened at one end, and positions were recorded at five 15-minute intervals. If the animals were indifferent to light, they would be in light 57 percent of the time (given the makeup of the tube). All four populations showed a significant negative response to light. From Vawter, Fong, and Culver (1987).

Population	Mean number of ommatidia	Percent of time in light	N
Davis Spring	28	20	38
Organ Spring	25	24	38
Organ Cave	3	46	27
Benedicts Cave	4	37	37

caves is accompanied by an increase in olfactory-lobe size (Fig. 3.4). The olfactory lobe is on average 25 percent larger in Organ Cave individuals than in Organ Spring individuals (Fig. 3.6). Antennae are also enlarged in cave populations. Holsinger and Culver (1970) found that the length of the first antenna was about 25–45 percent greater in a series of *tenuipes* populations than in spring populations, after taking into account variation in overall size of the animal (Fig. 3.7).

Although we have done no experiments on olfactory sensitivity, it is likely that cave populations, with longer antennae and larger olfactory lobes, are more sensitive to olfactory cues. It is certainly true that as a direct consequence of longer antennae, the radius of tactile sense is greater in cave populations.

A final consistent feature shared by cave populations is a large overall body size. While not all spring populations are smaller in body size (see Fig. 3.7), most are. Although large size appears to be a troglomorphic trait because of its parallel occurrence in independently derived populations, we believe that it is an instance of homoplasy but not troglomorphy. Spring populations of *G. minus* subject to fish predation, especially from sculpins *(Cottus)*, have a smaller body size at maturity than do populations in fish-free springs or in caves.

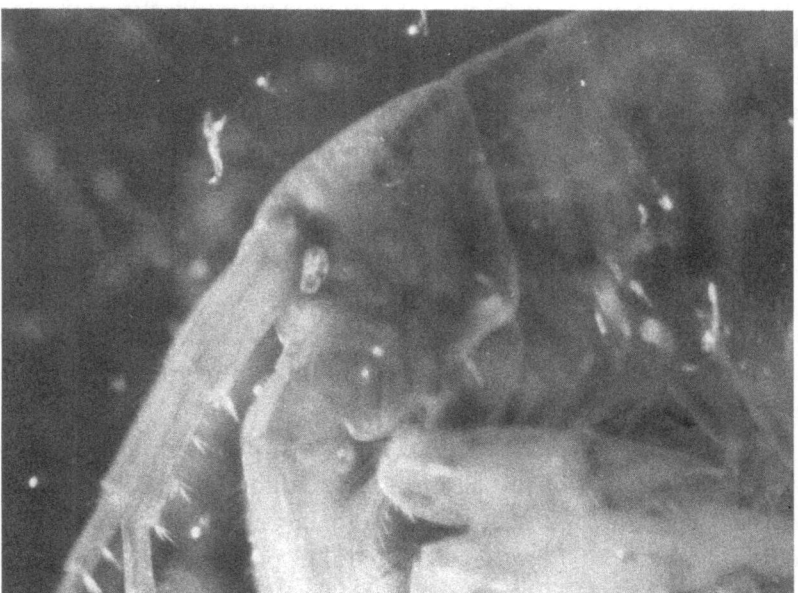

Fig. 3.2. The head region of *Gammarus minus* from Organ Cave resurgence *(top)* and Organ Cave *(bottom)*. Length of the first visible segment of the first antenna of the specimen from the resurgence is 0.4 mm and 0.5 mm for the Organ Cave individual.

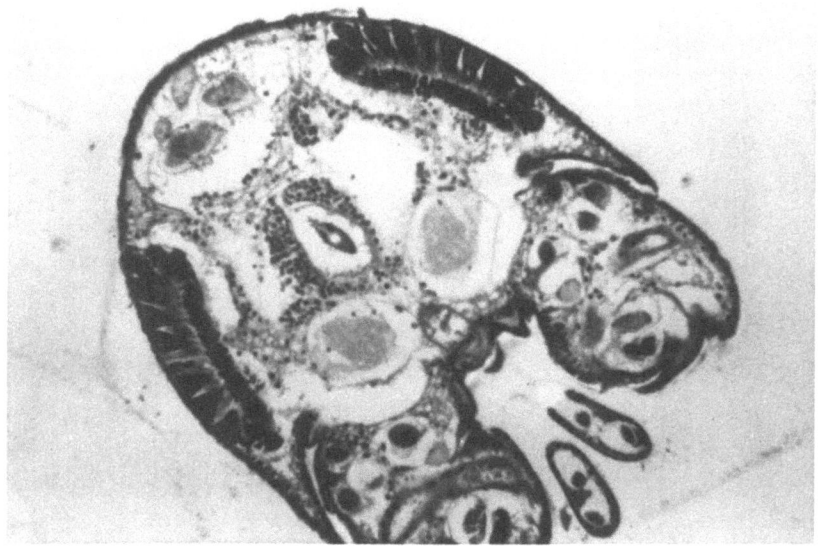

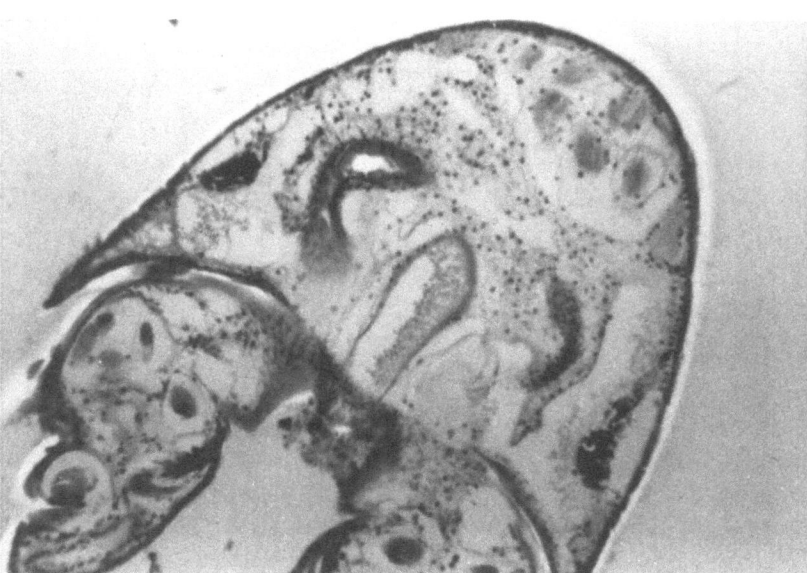

Fig. 3.3. Cross sections of the head of *G. minus* through the compound eyes. At top is of a typical spring individual from Davis Spring, West Virginia, and the bottom photograph is an individual from Organ Cave, West Virginia, displaying typical features for cave-dwellers. Both were stained with toludine blue and counter-stained with eosin. The width of the section between the eyes is 0.8 mm for the spring specimen and 1.2 mm for the cave specimen.

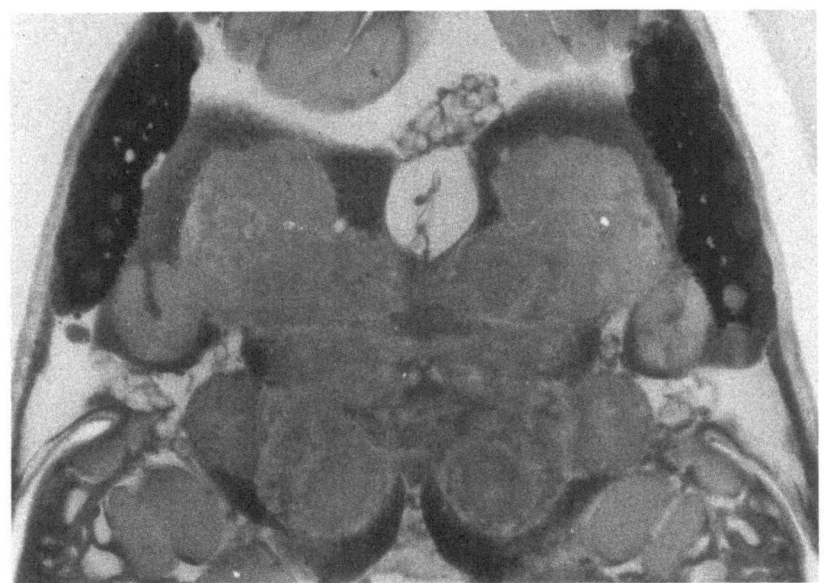

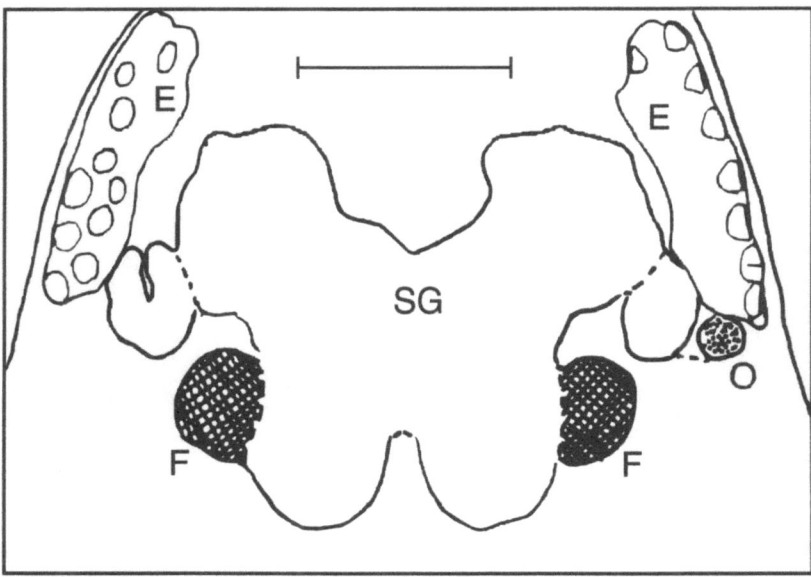

Fig. 3.4. Cross sections through the central nervous system at the level of the eyes of a *G. minus* from Organ Cave spring *(this page)* and a troglomorphic *G. minus* from Organ Cave *(facing page)*. *E* = eye, *O* = optic ganglion, *F* = olfactory lobe, and *SG* = supraesophageal ganglion. The optic ganglion is not visible in the specimen from Organ Cave. Scale bar = 0.2 mm.

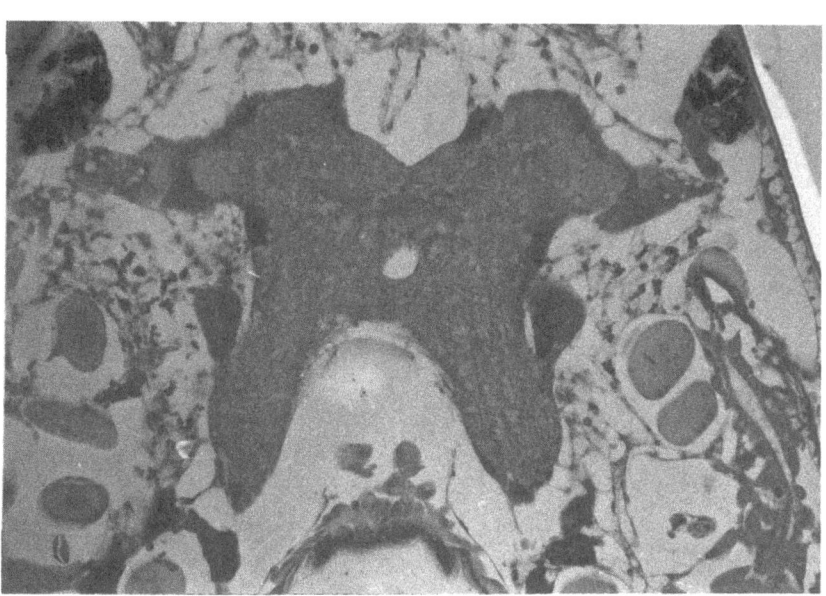

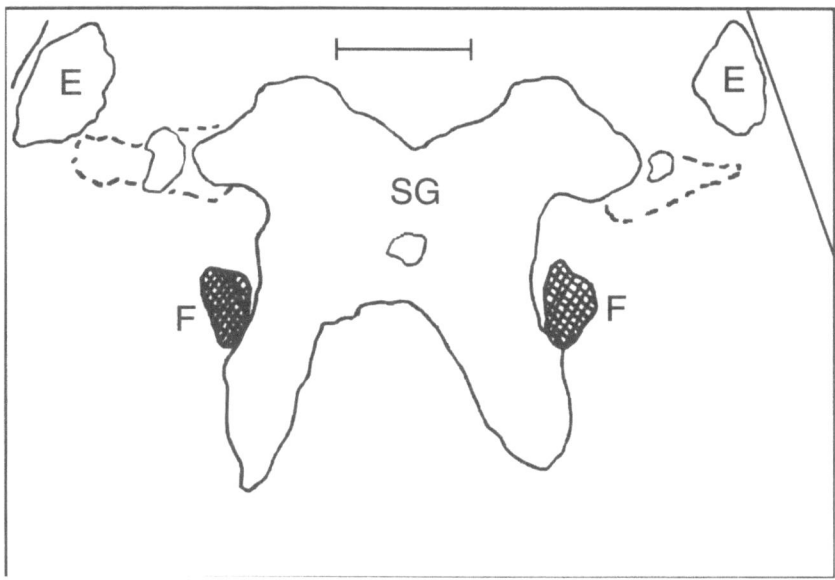

Fig. 3.4 (continued).

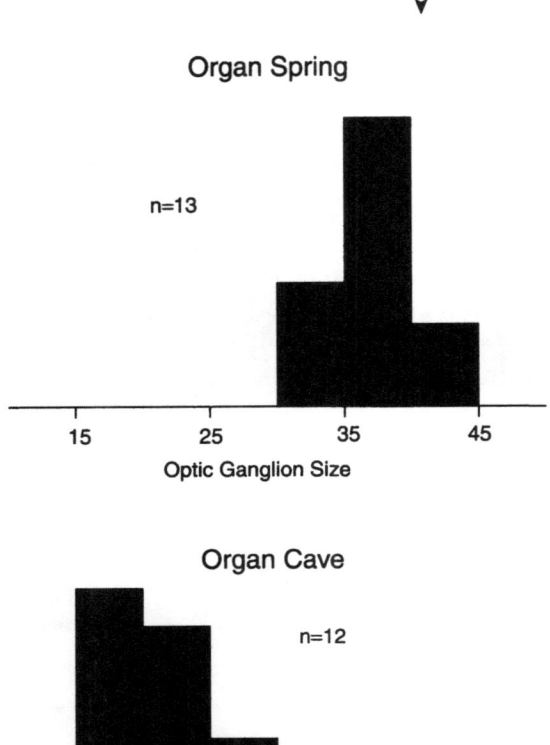

Fig. 3.5. Histograms of the size of the optic ganglion (400 units = 1 mm) in *G. minus* from Organ Spring and Organ Cave. Optic ganglia are significantly smaller in Organ Cave (Mann-Whitney U-test, $p < .01$).

Overview of the *Gammarus minus* Model

Our goal is to explain the causal mechanisms that produce the differences between cave-dwelling and surface-dwelling organisms. We will focus on morphological differences between cave and spring populations, the genetic basis of these differences, the repeated nature of these differences, and the geographic extent of the pattern.

Gammarus minus as a Model Organism

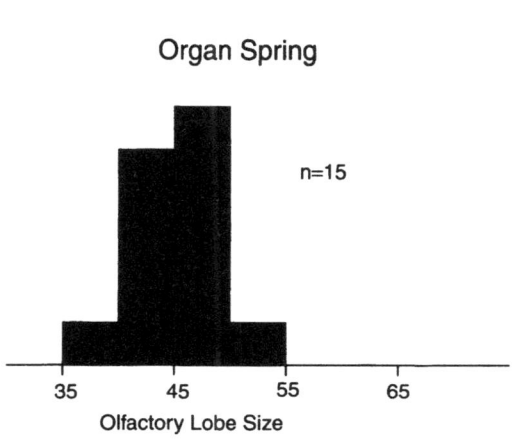

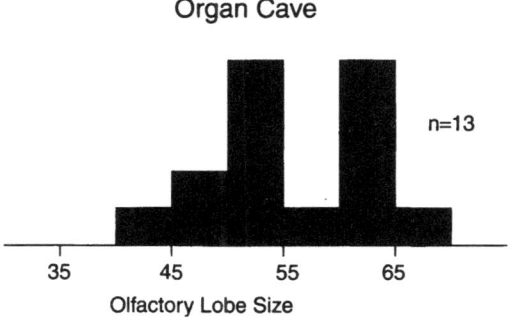

Fig. 3.6. Histograms of the size of the olfactory lobe (400 units = 1 mm) in *G. minus* from Organ Spring and Organ Cave. Olfactory lobes are significantly larger in Organ Cave (Mann-Whitney U-test, $p < .05$).

We begin by considering the ecology of *G. minus* in the context of environmental differences between spring and cave habitats. Next we consider the patterns, on a regional as well as on highly localized scales, of both morphological and genetic variation in *G. minus* in relation to the physical geography of caves and karst. We also summarize the evidence for a genetic basis for morphological features as well as evidence for natural selection producing morphological differences. Using this information, we offer our scenario for the evolu-

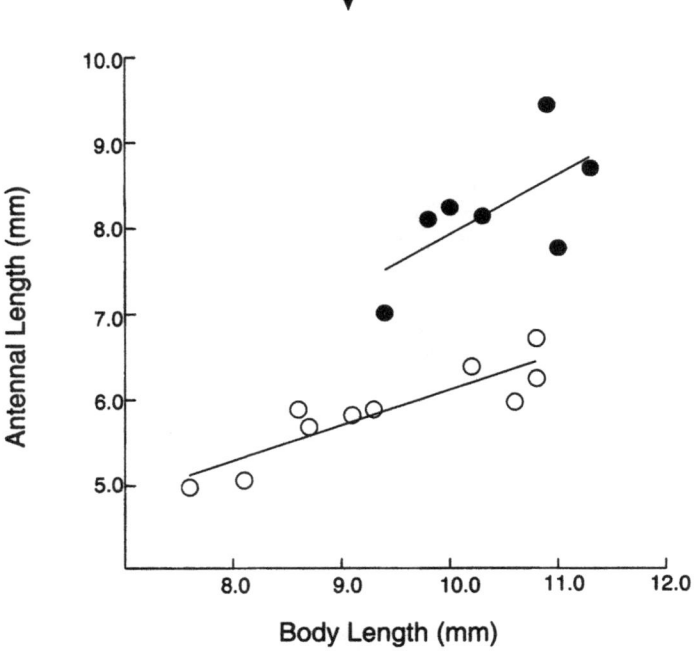

Fig. 3.7. Regression of mean body length against mean first antenna length for ten spring populations *(open circles)* and seven highly modified *tenuipes* populations *(closed circles)* of *G. minus*. Modified from Culver (1982).

tion of *G. minus*. In the concluding chapter we argue that the *G. minus* model is an empirical paradigm for the roles of common descent and natural selection in evolution and adaptation.

Summary

Gammarus minus occurs in great abundance in both surface and cave habitats. Ancestral character states of features that in cave organisms have become troglomorphic are retained in the surface populations. Cave-dwelling *G. minus* populations have smaller eyes and longer antennae than are found in surface populations, and they have smaller optic ganglia and larger olfactory lobes in the central nervous system. Correspondingly, cave populations show weaker photophobic behavior relative to spring populations.

Troglomorphic characters show considerable variation within and among populations. The high degree of morphological variation and the presence of multiple troglomorphic cave populations make *Gammarus minus* an excellent subject for evolutionary analyses.

Selected References

Holsinger, J. R. 1972. *The freshwater amphipod crustaceans (Gammaridae) of North America.* Biota of Freshwater Ecosystems Identification Manual No. 5. Washington, D.C.: Environmental Protection Agency. A thorough description of *Gammarus* morphology in general and *G. minus* morphology in particular.

Kosswig, C., and L. Kosswig. 1940. Die Variabilität bei *Asellus aquaticus* unter besonderer Berucksichtigung der variabilität in isolierten unter- und oberirdischen Populationen. *Revue de Facultie des Sciences* (Istanbul), ser. B, 5:1–55. The classic study of morphological variation in cave populations of *Asellus aquaticus*.

Mitchell, R. W., W. H. Russell, and W. R. Elliott. 1977. *Mexican eyeless characin fishes, genus* Astyanax: *Environment, distribution and evolution.* Special Publication No. 12. Lubbock: Texas Tech University Museum. The classic study of morphological variation in cave populations of *Astyanax*.

Poulson, T. L., and W. B. White. 1969. The cave environment. *Science* 165:971–981. A discussion of caves as model systems not only for evolutionary questions but for ecological and mineralogical ones as well.

4 The Ecological Theater

Having introduced the main character, *Gammarus minus*, in Chapter 3, we now set the stage for our story—to use G. E. Hutchinson's words, the "ecological theater" in which it performs. The species is found in running waters, both aboveground in springs and resurgences, and in cave streams. Lotic environments may be quite variable, and in this chapter we review the physical and biotic characteristics of both spring and cave habitats. We then consider the life cycle of *G. minus*, its trophic relationships, its biotic interactions, and its basic physiology.

Springs

Gammarus minus will tolerate only low temperatures, a feature it shares with many other species of *Gammarus* occupying lotic habitats (Marchant 1981). It is rarely found where the water temperature exceeds 15°C, and it is absent where the temperature exceeds 20°C for some duration in the summer. In a study of eight spring-dwelling *G. minus* populations in southern and northeastern West Virginia, Man (1991) obtained year-round bimonthly temperature records for each spring. The annual average temperatures for all eight springs were within a range of 9–12°C, and the temperatures of each spring fluctuated over an annual range of less than 2°C. Glazier, Horne, and Lehman (1992) reported temperature readings for ten springs in Pennsylvania with populations of *G. minus*. One spring was measured at 18°C, while the other nine were within 9.5–12.5°C. They also obtained year-round monthly temperature records for five of the springs, which showed annual and monthly fluctuations of 3°C or less. The temperature in another Pennsylvania spring harboring *G. minus* varied only from 9.1 to 11.0°C over two years (Gooch and Glazier 1991).

The habitats of *Gammarus minus* are characterized by hard, alkaline water. Glazier, Horne, and Lehman (1992) surveyed thirty-two springs in central Pennsylvania that showed a total range in pH from 4.6 to 7.7, and in conductivity from 14 to 411 µS/cm, although monthly measurements of pH and conductivity taken at each of five of the springs showed little temporal variation. They found that *G. minus* was present in all but two of the twenty-three springs with conductivity higher than 30 µS/cm and pH values of 6 or higher, and that *G. minus* was absent from ten of twelve springs with conductivity of 50 µS/cm or lower and absent from all nine springs with pH of less than 6 (Fig. 4.1). Furthermore, the population density of *G. minus* was positively correlated with pH and conductivity, as well as with concentrations of calcium ion, total calcium and magnesium ions, and calcium carbonate.

Shuster and White (1971) suggested that the extent of annual variability, rather than the absolute values, of physical and chemical parameters of a spring is a good indicator of the type of aquifer feeding the spring. They found that water-hardness readings over one year showed coefficients of variation from 10 to 24 percent in springs fed by conduit flow, but less than 5 percent in springs fed by diffuse flow. The water temperature also fluctuated over an annual range of 6–8°C in conduit-flow springs but was remarkably constant in diffuse-flow springs. Furthermore, springs with little or no fluctuations in temperature had higher total hardness and calcium ion and bicarbonate ion concentrations as well as higher pH. Not surprisingly, diffuse-flow springs were near saturation with respect to calcium, whereas conduit springs were undersaturated by factors of 2 to 5. *G. minus* is found in both diffuse and conduit springs, but it may be missing from some conduit springs because of large fluctuations in temperature, discharge, conductivity, or the like. For example, *G. minus* has not been found in the resurgence of Culverson Creek Cave in West Virginia

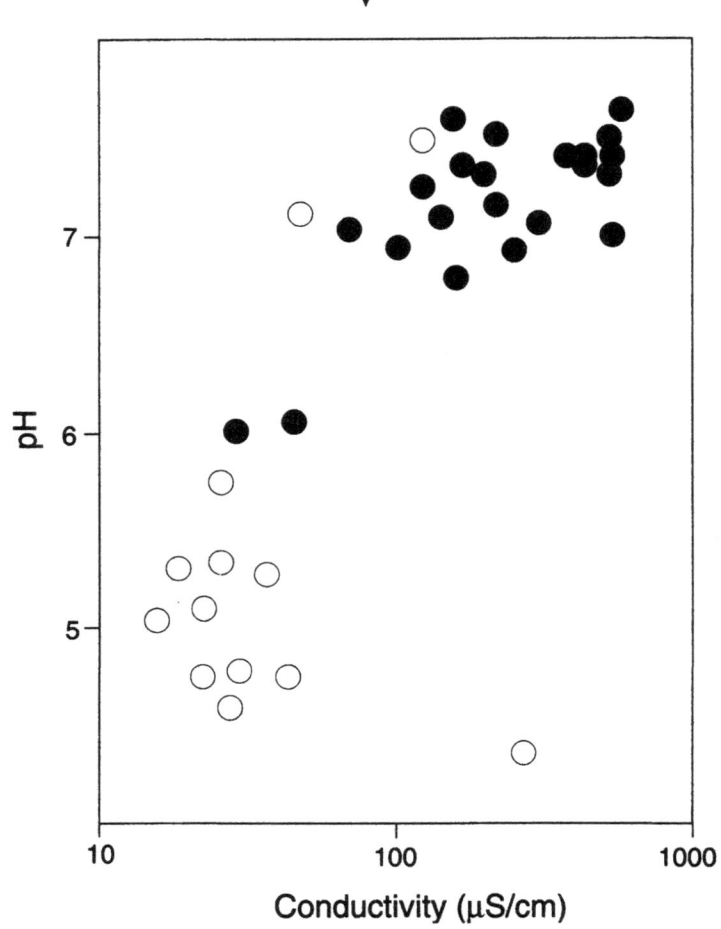

Fig. 4.1. Conductivity and pH of thirty-two springs in central Pennsylvania. A solid circle indicates *G. minus* was present; an open circle indicates *G. minus* was absent. Modified from Glazier, Horne, and Lehman (1992).

(see Jones 1973). Other conduit springs where *G. minus* is found, such as the ones studied by Man (1991), appear less variable than the conduit springs studied by Shuster and White. This pattern may be explained by the concentration of calcium near saturation levels in such springs, because rapid uptake of calcium is critical for *G. minus* during molting (see below).

Caves

Students of the biology of subterranean faunas have long stressed the physical and chemical stability of subsurface habitats (Barr and Kuehne 1971, Poulson and White 1969, Culver 1982). Of course there is one physical characteristic for which there is no doubt about increased stability—the amount of light. Populations of *G. minus* in cave streams encounter no light. However, the surface-water habitats of *Gammarus minus*, springs and spring runs, are not typical of aboveground watercourses; they can best be thought of as ecotones, or boundary zones between surface and subterranean waters (Gibert et al. 1990). In fact, by most chemical and physical measures (except for incident light), cave streams are less stable than the springs through which the cave stream water resurges.

Populations of *G. minus* living in cave streams may experience greater fluctuations in physical and chemical parameters than their spring-dwelling counterparts encounter. For example, the temperature of a cave stream may drop drastically from an influx of melting snow in the spring, or it may rise suddenly from a flush of warmer water from a summer storm. The severity of such disturbances on the temperature of a particular cave stream will depend on the proximity of the cave stream to insurgences, sites where surface water enters the cave system. By the time the water resurges at a spring, it has been underground longer and is closer to equilibration with the surrounding rock. The physical and chemical regimes of a spring are thus more constant than those of cave streams, because the effects of any thermal disturbance originating at insurgences would be damped or diluted by the time the stream water reached the spring.

A short-term profile of both habitats for *G. minus*—the cave stream and the aboveground spring—is available for two streams in Organ Cave and the resurgence spring of Organ Cave (Zeit 1993). All three sites support dense populations of *G. minus*. As measured by temperature, recorded at 15-minute intervals for 49 days during the summer of 1993, both cave streams were more variable than the spring. In spite of a diurnal temperature cycle at the spring, its coefficient of variation was smaller (Table 4.1).

Table 4.1. Variability of temperature (°C) in two cave streams in Organ Cave and the resurgence of Organ Cave. Organ Stream is fed largely by percolation and the 1812 Stream is fed by direct input from a surface stream. Data are based on 4,601 measurements taken at 15-minute intervals during late summer of 1992 (Julian days 219 to 267).

Location	Mean	Maximum	Minimum	Coefficient of variation
1812 Stream	11.5	11.7	11.2	0.027
Organ Stream	8.9	9.0	8.7	0.011
Resurgence	8.4	9.0	8.2	0.008

Storms had different effects on relative water levels in the two cave streams (Fig. 4.2). In one cave stream (1812 Stream) with an open input from the surface, water rose quickly and fell quickly following a rain. In Organ Stream, with diffuse input from the surface, water level rose more slowly and fell very slowly. The spring showed a bimodal response that lagged behind the effects experienced in the two cave streams by several hours. Thus, the spring is less responsive to fluctuations caused by storms, and the response represents an averaging of those found in the subterranean tributaries.

Biotic Environments of *Gammarus minus*

Relative to other freshwater habitats, neither springs nor caves contain a particularly diverse fauna. The macroscopic fauna of one resurgence and three caves are listed in Table 4.2. The faunal variety of Organ Cave resurgence is typical for springs in the area and for cold-water karst springs in general (see Minckley and Cole 1963). The numbers of species found in the three cave streams listed in Table 4.2 (six to eight) are also typical (Culver 1970).

Besides reduced diversity, cave stream faunas also have less trophic complexity. Fewer large predators are found in cave streams than in springs. The salamander *Gyrinophilus porphyriticus* is the only large predator regularly present in cave streams. Although this predator

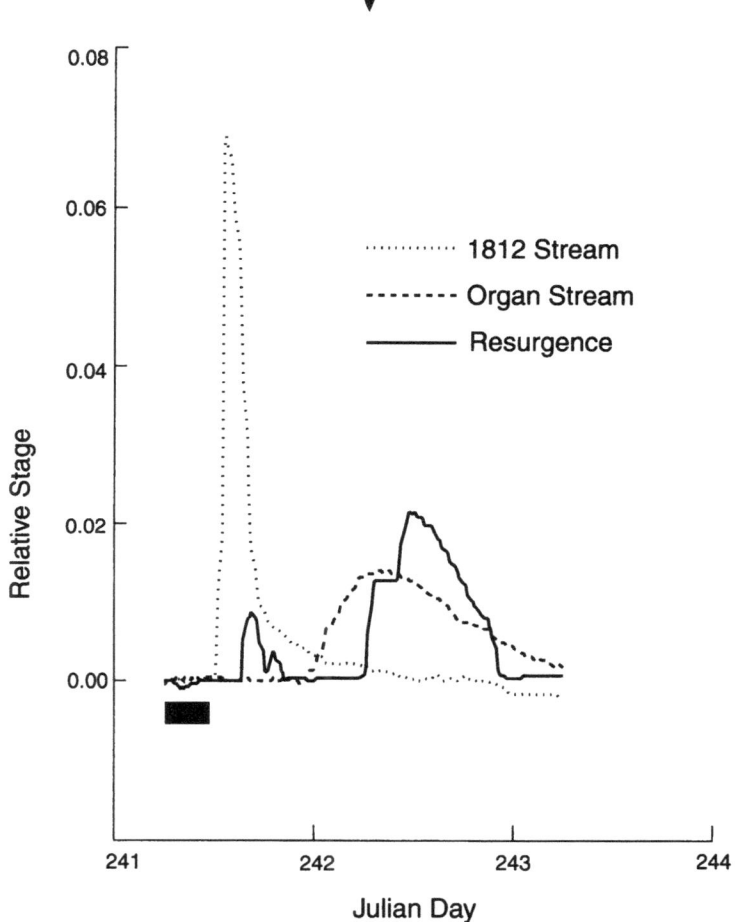

Fig. 4.2. Response of two streams in Organ Cave and of the Organ Cave resurgence to a rainfall of approximately 1 cm. For each site, the water level (a correlate of discharge) is scaled so that the total area under the curve is identical (= 1). Thus, only information about shape of the response is available. The solid bar indicates the duration of the rain.

has a major effect on the microdistribution of amphipods and isopods in cave streams, it rarely affects species composition itself (Culver 1975).

An amphipod population invading a cave faces reduced numbers of predators, but it does not necessarily face reduced numbers of

Table 4.2. Macroscopic fauna of one spring (Organ Cave resurgence, ORS) and three caves (Organ Cave, ORG; The Hole, HOL; and Benedicts Cave, BEN), all of which contain large *Gammarus minus* populations. Transients and accidentals have been excluded.

Taxa	ORS	ORG	HOL	BEN
PLATYHELMINTHES: TURBELLARIA				
Order Tricladida				
Phagocata gracilis	X		X	
Macrocotyla hoffmasteri				X
MOLLUSCA: GASTROPODA				
Order Limnophila				
Lymnaea sp.	X			
Physa sp.	X			
Order Mesogastropoda				
Hydrobiidae sp.	X			
Fontigens tartarea		X		
Fontigens turritella			X	
ARTHROPODA: CRUSTACEA				
Order Amphipoda				
Crangonyx gracilis	X	X		X
Gammarus minus	X	X	X	X
Stygobromus emarginatus		X	X	X
Stygobromus spinatus		X	X	X
Order Isopoda				
Caecidotea holsingeri		X	X	X
Caecidotea scrupulosus	X			X

competitors. Aside from *G. minus*, between three and five amphipod and isopod species are present in cave streams while only two other amphipod and isopod species are present in the spring (Table 4.2).

Population Dynamics and Life Cycle of *Gammarus minus*

In general, both spring populations and morphologically modified cave populations of *G. minus* are quite large, usually well over 10,000 individuals. In one stream in Organ Cave, densities averaged over 100

Table 4.2 (continued).

Taxa	ORS	ORG	HOL	BEN
ARTHROPODA: UNIRAMIA				
Order Coleoptera				
Haliplus sp.	X			
Order Diptera				
Dixa sp.	X			
Simulium sp.	X			
Order Ephemeroptera				
Baetis sp.	X			
Order Plecoptera				
Leuctra sp.	X			
Taeniopteryx sp.	X			
Order Trichoptera				
Brachycentrus sp.	X			
Lepidostoma sp.	X			
Molanna sp.	X			
CHORDATA: OSTEICHTHYES				
Order Cypriniformes				
Cottus bairdii	X			
Order Perciformes				
Lepomis sp.	X			
CHORDATA: AMPHIBIA				
Order Caudata				
Gyrinophilus porphyriticus		X		X
Eurycea lucifuga		X		

individuals/m^2 (Culver et al. in press). This stream is 1 km long and 0.5 m wide, resulting in a population in that stream alone of approximately 5×10^4. Culver (1971) recorded densities averaging more than 200/m^2 in a 5 km stream in Benedicts Cave, a population of 10^6. More typically, densities are 10/m^2 in cave streams. Given that cave streams with morphologically modified *G. minus* populations are usually many kilometers in length, population sizes are large.

Many spring populations are also large, including those in the resurgences of the large cave systems we have studied. Many springs are difficult to sample quantitatively, since the habitat is a complex three-dimensional matrix of considerable depth. Jones (1990) does provide information on a short spring run (Dickson Spring). Here *G. minus* is found in gravels and in watercress in a shallow stream at densities averaging nearly 200/m^2 with at least 1,000 m^2 of habitat. The population in Dickson Spring is thus likely to be approximately 2×10^5. There are of course smaller populations of *G. minus*. Many apparently unmodified cave populations of *G. minus* are quite small, but we know of no cave-modified population of *G. minus* that is small.

Gammarus minus follows a life cycle that is typical of freshwater gammarid amphipods (see Hynes 1955). How *Gammarus* recognize mates is not clearly understood. Dahl, Emanuelson, and von Mecklenberg (1970) suggested that the process is mediated by pheromones released by females and detected through the calceoli, which are antennular organs in males. Read and Williams (1990), however, showed that the calceoli in *G. pseudolimnaeus* cannot be pheromone receptors because they are not innervated, and that males with ablated calceoli performed equally well as control males in the ability to locate females and form pairs. Godfrey, Holsinger, and Carson (1988) also suggest that the calceoli of *G. minus* are not innervated. Whatever sensory structures are involved, *Gammarus* males can readily locate females, especially ones closer to molting and presumably more sexually receptive (Hartnoll and Smith 1978, Dunham, Alexander, and Hurshman 1986).

Prior to copulation, sexually mature males and females form precopula pairs, with the larger male (8–12 mm body length) carrying the smaller female (5–9 mm body length) by grasping the female with the gnathopods and pressing her dorsum to his ventral side (Borowsky 1984). During precopula the female is passive and all activities are conducted by the male. In the laboratory the precopula stage lasts up to two weeks. Copulation takes place after the female releases eggs into a brood pouch. The male then turns her over and fertilizes the eggs by depositing packets of sperm into the brood pouch. Presumably, the female must molt just prior to releasing her eggs into the brood pouch (Hynes 1955), and this may serve as a physical cue, a supplement to possible chemical cues (Dahl, Emanuelson, and von Mecklenberg 1970), for copulation to proceed.

The pairs separate soon after copulation. Ovigerous females carrying fertilized eggs are readily recognizable by the mass of dark yolk visible through the transparent brood pouch. The eggs develop and young amphipods then hatch within the pouch. Ovigerous females carrying eggs in late developmental stages or newly hatched young are recognizable by the yellowish mass in the bulging brood pouch. A few newly hatched young sometimes can manage to squeeze out of the brood pouch by themselves, but normally the entire brood is released after the female molts.

The newly released young amphipods are independent miniature versions of the adults, at about 1 mm total body length. Females of *G. minus* can produce multiple broods within a season. While embryos are developing in the brood pouch, new eggs are maturing in the ovary. When the first brood is about to hatch, the female is again receptive and pairs with a male. We have observed that the females in many precopula pairs from both cave and spring populations of *G. minus* are carrying eggs in very late developmental stages or even newly hatched young. When brought into the laboratory, these pairs will copulate immediately after the female molts and releases her previous brood. Read and Williams (1990) also observed that among collections of *G. pseudolimnaeus* from two Ontario streams, 62 percent of all females paired in precopula were already carrying developing broods.

In spring populations breeding activity usually peaks during the winter months, when up to 70 percent of adults are in precopula pairs (less than 5 percent are paired when breeding activity is at the lowest point). The population in Falls Ravine Creek, Pennsylvania, studied by Kostalos (1979), showed the highest breeding activity from December through February. The proportion of females carrying developing eggs increased sharply from 5 percent in November to over 40 percent in December and continued to increase to a peak of over 80 percent in May. The proportion of ovigerous females in the population then declined, and almost no ovigerous females were observed in October. A typical individual in this population thus develops in the maternal brood pouch during late winter to early spring and is released sometime from late March to May. Its probability of survival is very low; there is a high mortality rate for the newborn and juvenile size classes. If it survives it grows to sexual maturity and mates between December and February. If it is female, it releases one or more broods during March through May. It then dies by November. The life cycles of eight spring populations of *G. minus* in West Virginia studied by Man (1992), with some slight variation in timing of events, are remarkably similar to the pattern just summarized.

Life-cycle patterns of cave populations are less clear. In some cases, populations appear to have a mid-winter peak in precopulation pairing, and the young are hatched in spring (Culver 1971, Jones 1990). In other cases, no pattern was discernible. The most that can be said is that, to the extent that there is a cycle, it is similar to that observed in spring populations.

The fact that distinct peaks in breeding activity are observed in both spring and cave populations raises the interesting question of possible environmental cues. Steele and Steele (1986) showed that in *G. lawrencianus* and *G. setosus* from intertidal and estuarine habitats in Newfoundland, reproductive cycles are controlled by endogenous rhythms entrained by photoperiod. Photoperiod likewise may entrain the reproductive cycles in spring-dwelling populations of *G. minus*, but it certainly has no direct effect on cave-dwelling populations.

Because breeding activities usually peak between December and February, Kostalos (1979) suggested that decreasing temperatures may be the cue for onset of reproduction. We were also able to induce an increase in precopula pairs in laboratory cultures of cave and spring populations by dropping the incubator temperature from 10°C to 4°C. This decrease is greater than the annual temperature fluctuations of many *G. minus* habitats, but a smaller drop may also induce the formation of precopula pairs.

Another environmental stimulus may be an increase in the influx of organic detritus. The accumulation of leaf litter during the fall should increase the amount of detritus washed into cave streams by winter storms. An increase in available food is a plausible cue for the onset of reproduction in cave populations.

Brood size of *G. minus*, usually measured as the number of developing embryos in the brood pouch, shows much variation among populations, but it generally also shows a positive covariation with maternal body size. Glazier, Horn, and Lehman (1992) showed that, for ten spring populations from Pennsylvania, mean brood sizes ranged from 4.2 to 9.7 and mean maternal body lengths ranged from 5.2 mm to 7.1 mm, and that both variables differed significantly among the populations (Fig. 4.3). Among these populations, maternal dry mass was a good predictor of brood dry mass ($r^2 = 0.90$) and maternal body

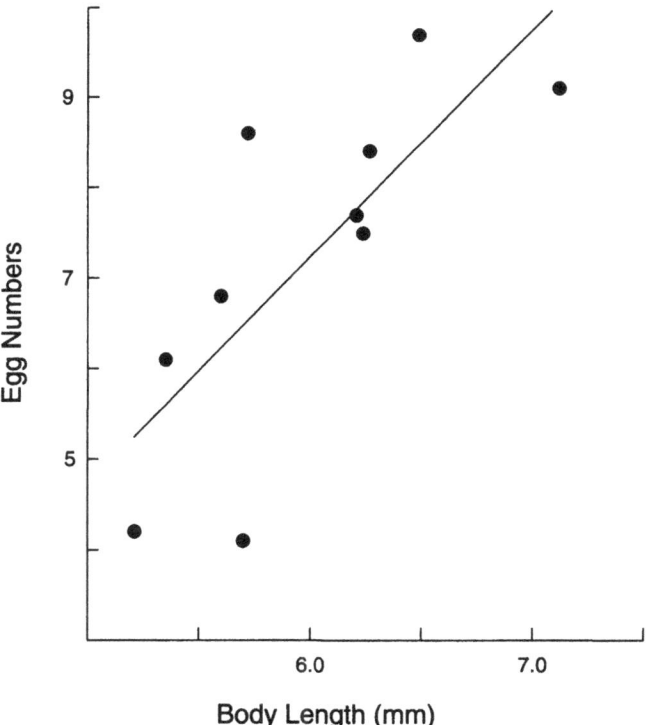

Fig. 4.3. Relationship between mean female body length and mean egg number for ten *G. minus* poulations from Pennsylvania springs. The linear regression accounts for 57 percent of the variance in egg number. Data from Glazier, Horne, and Lehman (1992).

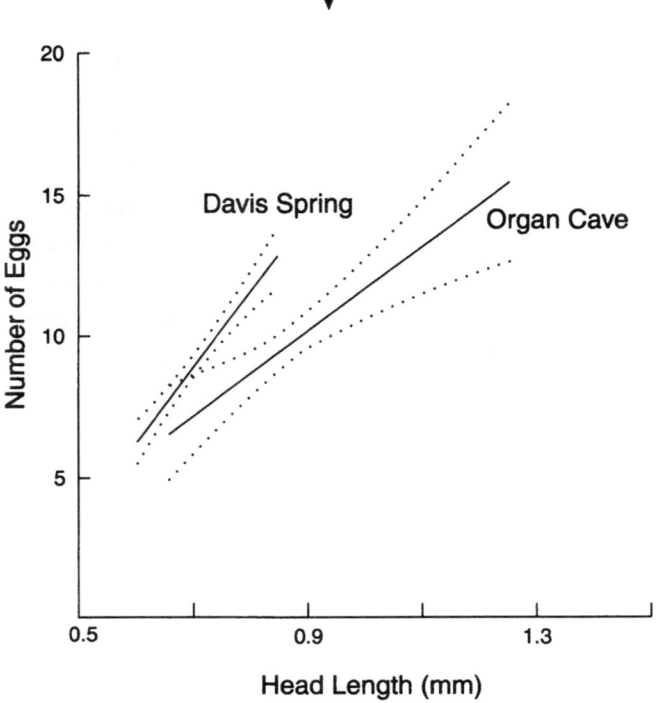

Fig. 4.4. Relationship between female head length and egg number of *G. minus* from Davis Spring and Organ Cave. Dotted lines are 95 percent confidence intervals. Over 250 individuals from each population were measured.

length was a good predictor of brood number ($r^2 = 0.57$). Whether the same relationship exists between maternal body size and brood size within populations, however, was not reported.

Brood size does increase with head length for both Organ Cave and Davis Spring populations (Fig. 4.4). In addition, number of eggs was greater and increased faster with greater head length in the Davis Spring population.

Trophic Relationships and Biotic Interactions

In common with other species of amphipods in running-water habitats, *G. minus* is a consumer of organic detritus, particularly allochthonous plant material such as fallen leaves, though it derives most of its nutrition from the bacteria and especially fungi associated with the detritus (Hargrave 1970, Kaushik and Hynes 1971, Bärlocher and Kendrick 1973, Moore 1975, Marchant 1981). In our laboratories populations of *G. minus* from both spring and cave habitats rapidly skeletonized fallen elm leaves *(Ulmus americanus)* that were first softened by boiling for several hours. Laboratory feeding experiments by Kostalos and Seymour (1976) showed that *G. minus* exhibited significantly higher survivorship and higher fecundity on all fungal-enrichment treatments (Table 4.3). Bacterial enrichment resulted in lowered survival but had ambiguous effects on fecundity. Most strikingly, almost no adults survived and none reproduced when leaf detritus had been sterilized or the microflora killed (Table 4.3).

Although fallen leaves are readily available in springs, their supply in most cave streams is more limited. Even in cave stream habitats

Table 4.3. Relative survival and fecundity of *G. minus* after 10 weeks on various diets. The results of several experiments of Kostalos and Seymour (1976) are set to the same scale by setting data for the control group equal to 100 and dividing all experimental values by the actual control values and multiplying by 100. From Culver (1985).

Diet	Survival	Fecundity
I. Control: Leaf detritus and associated microflora	100	100
II. Fungal enrichment		
A. Bacteria killed,[a] detritus	133	151
B. Fungi only, no detritus	118	115
C. Fungi added to detritus	106	113
III. Bacterial enrichment		
A. Bacteria only, no detritus	63	136
B. Bacteria added to detritus	76	53
IV. Detritus alone		
A. Sterile detritus	7	0
B. Detritus with microflora killed and removed	7	0

a. Bacterial reduction causes an increase in fungus.

with open inputs, such as 1812 Stream in Organ Cave (Fig. 4.2), input of leaves is largely the result of rains. It is unlikely that the high population density of *G. minus* observed in some cave streams can subsist solely on the limited periodic influx of leaves and the associated microflora. Although *G. minus* in cave streams is associated with plant detritus, they may obtain much of their nutrition from cave mud (Culver 1985). Uptake of nutrients onto stream sediments is a well-documented process; it may occur through abiotic adsorption onto the substrate or through absorption by microorganisms associated with the substrate. Wallis (1981) calculated that the average rate of uptake of dissolved organic matter (DOM) by sediments in a spring-fed first-order stream in Alberta was about 200 $mg/m^2/h$. Elwood et al. (1981) showed that most of the uptake of phosphorus by the substrate in Walker Branch, Tennessee, is through microbes associated with the sediment.

G. minus may also feed on the bacteria and other microorganisms associated with the surface film in cave pools. We have frequently observed *G. minus* individuals oriented with the ventral side upward, lying at the edge of pools or clinging just below the water line on the side of rocks, and rapidly working its mouthparts right at the surface of the water. Carol Haley (pers. comm.) obtained evidence from scanning electron microscopy indicating that mouthparts of individuals from cave habitats are much more setose, or bristly, than those of spring-dwelling individuals, and suggested that the setae may aid in filter feeding. We have also observed *G. minus* individuals in cave streams feeding on remnants of the oligochaete *Stylodrilus beattiei* and the amphipods *Stygobromus emarginatus* and *S. spinatus*. However, whether *G. minus* was acting as a scavenger or as an outright predator in such incidences is unknown.

Under laboratory conditions, at least, *G. minus* is an efficient predator on *Caecidotea holsingeri*, an isopod common in many cave streams where *G. minus* occurs (Culver, Fong, and Jernigan 1991). It is apparent, nonetheless, that in cave habitats *G. minus* subsists mainly on nutrients extracted from the sediment or from microorganisms associated with the surface film and will feed opportunistically on detritus and other available organisms. A broad diet is characteristic of organisms adapted to the aquatic cave environment (Culver 1985).

In general, *Gammarus* plays a significant role in the processing of organic detritus in stream ecosystems. Marchant and Hynes (1981) obtained field estimates of the feeding rate of *G. pseudolimnaeus* in a stream in Ontario. They calculated that this species alone consumed 800 kg/ha/yr, accounting for about 16 percent of the average allochthonous input into the stream. Although *Gammarus* is capable of consuming a large quantity of detritus, the assimilation efficiency of a population is usually quite low, at about 10–20 percent (Bärlocher and Kendrick 1975; Marchant 1981). Therefore, 80–90 percent of the large amount of detritus consumed by *Gammarus* is egested as fecal pellets. Marchant (1981) suggested that this feature of *Gammarus* populations is important in the stream and spring ecosystems because of the significant quantities of finely chewed plant material released in the fecal pellets. The much increased surface area of these pellets will allow for more rapid colonization by bacteria and fungi, and they in turn serve as a major food source for other organisms as well as for newborn and juvenile amphipods not as well equipped to utilize large pieces of coarse detritus as food.

Because *Gammarus minus* is the dominant macroinvertebrate in terms of biomass and population density in many springs, reaching densities of more than $8000/m^2$ (Glazier, Horne, and Lehman 1992), it likely plays an important role in the initial release of nutrients bound in detritus and the subsequent spiraling of nutrients (Elwood et al. 1983) in springs. Likewise, the action of *G. minus* in cave streams may also affect the availability of nutrients. Microbial populations generally serve as nutrient sinks and as bottlenecks in energy flow (Macfadyen 1961). Utilization of cave sediment and of the surface film of cave pools by *G. minus* will promote the flow of energy tied up in microbial biomass, thus enhancing further heterotrophic production in the cave stream ecosystem.

Indeed, cave streams with large *G. minus* populations invariably have a diverse fauna of invertebrates, including isopods and other amphipods. In a perturbation study in Organ Cave (Culver, Fong, and Jernigan 1991), the amphipod *Stygobromus emarginatus* had higher survival rates when *G. minus* was in the vicinity. This is in spite of the fact that upon contact, the two species avoid each other. The positive commensal effect of *G. minus* on *S. emarginatus* is apparently the result of *S. emarginatus* feeding on fecal material of *G. minus*. *S. emarginatus* individuals showed a significant weight gain in the lab when their diet was supplemented with *G. minus* feces (Culver et al. 1991).

Gammarus also play a significant role in energy flow in streams and springs when they are taken as prey by fish, especially trout *(Salmo* and *Salvelinus)* and sculpins *(Cottus).* Waters and Hokenstrom (1980) estimated the annual production of *Gammarus pseudolimnaeus* in a Minnesota stream at 64–271 kg/ha, and Marchant (1981) suggested that at least 20 percent of the annual production in this species is lost to predation. Welton (1979) calculated that at least 60 percent of the annual production of *G. pulex* was lost to fish predation. Further, fish predation is demonstrated to be size-selective on larger individuals of *G. pseudolimnaeus* (Newman and Waters 1984) and another amphipod, *Hyalella azteca* (Milstead and Threlkeld 1986).

Although comparable estimates of annual production and predation rate for *G. minus* do not exist, evidence suggests that epigean populations of *G. minus* are subject to heavy fish predation. Results from a study by Holomuzki and Hoyle (1990) of a population inhabiting a second-order stream in Kentucky indicate that predation by the green sunfish *(Lepomis cyanellus)* and the creek chub *(Semotilus atromaculatus)* reduced the amphipod densities in the stream runs. Man (1992) showed that several spring-dwelling populations of *G. minus* in West Virginia are subjected to size-selective predation by the sculpin *Cottus bairdi,* resulting in significantly smaller adult body sizes in these populations than in populations from springs where sculpins are absent.

Cave-dwelling populations of *G. minus* are not subject to fish predation, but they are preyed upon by salamanders such as the larvae of *Gyrinophilus porphyriticus* and possibly by crayfish *Cambarus bartonii* and *C. nerterius,* which are frequently found in cave streams. The density of such potential predators in cave streams is generally quite low, however, so it is not likely that these predators have a significant effect on the amphipod populations (Culver 1975). Unlike spring-dwelling populations, which transfer a considerable amount of energy between trophic levels, cave-dwelling populations of *G. minus* may act as nutrient sinks themselves and thus serve as bottlenecks in energy flow in cave stream ecosystems.

Basic Physiology: Ionic Balance and Metabolic Rates

Gammarus physiology is affected by concentrations of critical ions. For example, the importance of calcium in the environment of *G. minus* is underscored by the fact that calcium levels in *Gammarus* change significantly during the molt cycle. A study of calcium regulation in *Gammarus pulex* by Wright (1980) showed that adult specimens can lose more than 40 percent of body calcium during a two- to three-day period prior to molting, and that an additional 54 percent of body calcium is lost with the exuviae upon molting. Thus, rapid uptake of calcium from the immediate environment by newly molted individuals is essential for survival.

The exclusion of *G. minus* from acidic springs is an interesting phenomenon. Glazier, Horne, and Lehman (1992) hypothesized that the absence of *G. minus* from acidic waters may result from physiological stress, such as osmotic imbalance; ionic imbalance problems, especially those associated with growth and molting, such as reduced rates of calcium uptake after molts; and the generally poor food quality.

Another physiological function affected by environment is metabolic rate. Lowered metabolic rates have been suggested as troglomorphic features (see Table 2.1), but experiments using *Gammarus minus* (Culver and Poulson 1971) actually showed the reverse (Fig. 4.5). Two cave populations had metabolic rates higher than those measured in a resurgence population and a karst-window population. The biological significance of the higher rates is unclear, but it is clear that there has been no evolution of reduced metabolic rate in

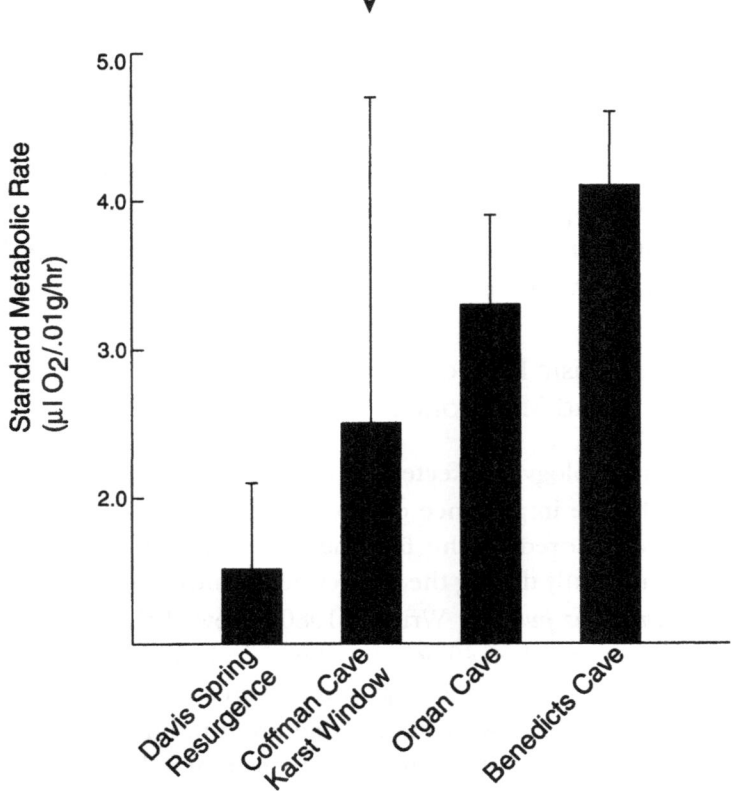

Fig. 4.5. Standard metabolic rates of four populations of *Gammarus minus*. Coffman's Cave is close to Higginbothams Cave (see Fig. 5.7). All but Organ Cave are in the same basin (see Chapter 5). Data from Culver and Poulson (1971).

cave populations, presumably because of the relatively rich detrital food base.

Summary

The surface and cave habitats of *Gammarus minus* are characterized by slightly alkaline waters with high conductivity and low temperatures. The species is absent from habitats with conductivity lower than 100 µS/cm and pH less than 6, likely because the combination of acidity and low concentration of calcium ions interferes with molting, moments of extreme physiological stress in the life of crustaceans in general. The thermal environment of *G. minus* is usually quite stable, averaging 10°C and fluctuating over an annual range of less than 3°C.

Although cave environments are generally considered more stable than surface habitats, our results show that, other than constant darkness, the physical and chemical regimes of cave streams inhabited by *G. minus* are actually more variable than those of springs. *G. minus* populations in caves have relatively more closely related competitors and fewer predators than spring populations face.

The life cycle of *G. minus* is typical of gammarids. Mature males and females form precopula pairs until the female deposits eggs into a brood pouch, the male then fertilizes the eggs, and the pair separates. The embryos develop and hatch within the brood pouch, and the young are released when the female molts. They develop into sexually mature adults in about 9 months in spring populations. Breeding activity in surface populations of *G. minus* is seasonal, peaking in the winter months. Breeding activity also shows mid-winter peaks in some cave populations, but not in others.

G. minus is a detritivore. In surface habitats it obtains its nutrition from the fungi and bacteria associated with plant detritus. Because *G. minus* exists in dense populations in springs, it plays a key role in the initial release of nutrients from the detritus in springs. Cave populations of *G. minus* feed opportunistically on plant detritus and other available organisms, and they may derive nutrition from bacteria and other microorganisms by processing cave sediments and surface film of cave pools.

Selected References

Culver, D. C., D. W. Fong, and R. W. Jernigan. 1991. Species interactions in cave stream communities: experimental results and microdistribution effects. *American Midland Naturalist* 126:364–379. A thorough study of biotic interactions of *G. minus* in caves.

Glazier, D. S., M. T. Horne, and M. E. Lehman. 1992. Abundance, body composition and reproductive output of *Gammarus minus* (Crustacea: Amphipoda) in ten cold springs differing in pH and ionic content. *Freshwater Biology* 28:149–163. A careful study of environmental conditions under which spring populations of *G. minus* occur.

Holomuzki, J. R., and J. D. Hoyle. 1990. Effect of predatory fish presence on habitat use and diel movement of the stream amphipod, *Gammarus minus. Freshwater Biology* 24:509–517. A field study of *G. minus* as prey.

Hynes, H. B. N. 1955. The reproductive cycle of some British freshwater Gammaridae. *Journal of Animal Ecology* 24:352–387. The classic source on the life cycles of gammarid amphipods.

Marchant, R. 1981. The ecology of *Gammarus* in running water. In *Perspectives in running water ecology*, ed. M. A. Lock and D. D. Williams, pp. 225–249. New York: Plenum Press. A general review of *Gammarus* ecology.

5 The Geography of *Gammarus minus*

In this chapter we place *Gammarus minus* in a geographic context. The spatial distribution of a species and the physical characteristics of its habitat have much to tell the evolutionary biologist. In the specific case of *Gammarus minus,* these details help us sort out the evidence for the evolutionary mechanisms (selective and otherwise) that result in the pattern of traits observed.

First we describe the historical biogeography of the genus, detailing its range, habitats, and affinities. For the geography of *G. minus* itself, we put special emphasis on the distribution of cave-dwelling populations and morphologically modified cave-dwelling populations. To explain these distribution patterns, we must take up a topic unfamiliar to most biologists, the physical geography of karst landscapes.

In Chapter 3 we described the major morphological differences between cave and spring populations. Here we turn to a more detailed analysis of variation in *G. minus*. We analyze data on variability of electrophoretically detectable proteins in the two areas where morphologically modified cave populations occur. Then we analyze geographic variation in morphology, especially in troglomorphic characters, again concentrating on those areas where both spring populations and highly modified cave populations occur. Finally, we compare and contrast genetic and morphological patterns of variation and ask what consequences they have for the evolutionary history of *G. minus*.

Biogeography of the Genus *Gammarus*

The genus *Gammarus* is widely distributed in freshwater habitats in the northern hemisphere. Barnard and Barnard (1983) list ninety freshwater species, of which twelve are in North America. There are an additional twenty marine species, derived from the freshwater species.

Gammarus is a very difficult genus taxonomically. In general it is a morphologically monotonous genus, with very minor characters separating most species (see Cole 1980, Pinkster 1983). Taxonomic problems are compounded by considerable microgeographic variation (Goedmakers 1980) and considerable within-population variation (Petre-Stroobants 1982). There is little consensus among amphipod taxonomists in the use of morphological criteria for establishing species designations. In spite of this taxonomic quagmire, there are some very clear distributional patterns in the genus.

First, *Gammarus* is predominantly found in higher latitudes in the northern hemisphere with, in Barnard and Barnard's phrase, "a preponderant affinity for cold water" (Barnard and Barnard 1983). It is widespread in lakes and streams in northern Europe (including Scandinavia) and Canada. In the United States, distribution of surface-dwelling species is largely limited to northern states (Fig. 5.1). The reasons for this northern Holarctic distribution are unclear. Amphipods are an old group, probably having evolved by the late Paleozoic, and *Gammarus* is a relatively old genus—it is well known from imprint fossils of Oligocene, Miocene, and Pleistocene age (Barnard and Barnard 1983). The absence of gammarids from warmer waters to the south may be historical, or it may be the result of their displacement by other groups, such as the amphipod genus *Crangonyx*.

Second, *Gammarus* seems to have a special affinity for springs, cave streams, and other subterranean habitats. In the recent compendium of aquatic subterranean organisms, *Stygofauna Mundi,* Stock (1986) lists ten species predominantly found in springs, and an additional nine predominantly found in caves. Most of the cave species have not completely lost their eyes but show varying levels of eye reduction and

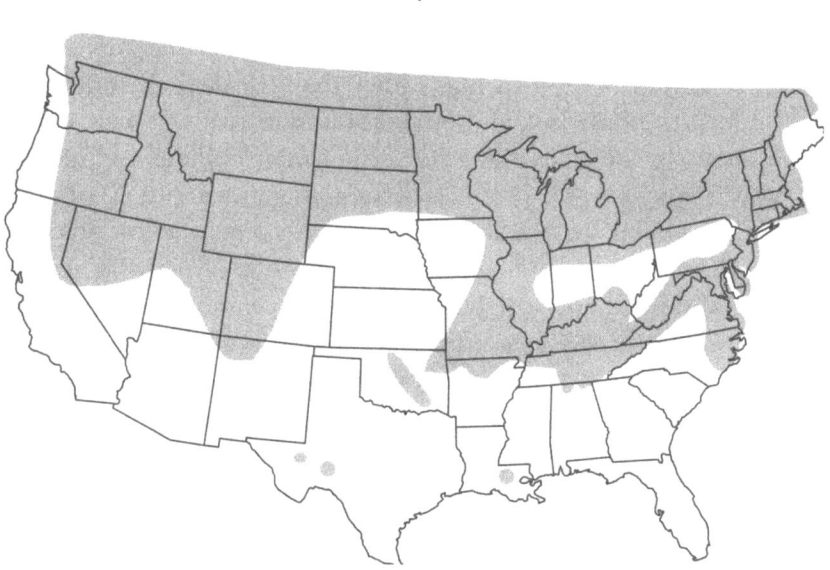

Fig. 5.1. Distribution of freshwater *Gammarus* in the United States. Distribution extends farther north than shown. From Barnard and Barnard (1983) and Delong (1992).

other modifications for subterranean life, as does *G. minus*. One well-known parallel to the pattern of *G. minus* is that of *G. pulex* in caves in Europe and Asia (Karaman and Pinkster 1977). *G. pulex* has invaded caves in several areas throughout its range, resulting in at least some genetic differentiation (Anders 1956), reduction of eyes, and additional morphological changes (Karaman and Pinkster 1977). Several apparently closely related genera (*Accubogammarus* and *Metohia* in the Balkan peninsula and *Zenkevitchia* in the Caucasus) have only eyeless species and are only found in caves. Nevertheless, compared with other groups of amphipods in caves, *Gammarus* is neither highly modified morphologically nor highly speciated. For example, there are over a hundred species of *Stygobromus* in North America, all in subterranean habitats and all without eyes or pigment (Holsinger 1986). Similar patterns obtain for the predominantly European genus *Niphargus* (Karaman and Ruffo 1986).

A third distributional pattern displayed by *Gammarus* is its affinity for caves and associated springs rather than for deep groundwater, the underflow of streams, and other habitats in sands. It is an affinity for "milieux perméables en grand" rather than "milieux perméables en petit" (Botosaneanu 1986). *Gammarus* in general and *G. minus* in particular inhabit surface waters, especially coldwater streams, and have invaded springs and subterranean streams directly. In this regard, *Gammarus* shares a habitat lineage in common with the widespread isopod *Asellus aquaticus* (Sket 1965), amblyopsid cave fish (Woods and Inger 1957), and some other amphipods such as *Crangonyx* (Holsinger 1986). On the other hand, there are many groups that primarily inhabit subsurface habitats other than caves—for example, crustacean groups such as Harpacticoida and Ostracoda (Danielopol and Rouch 1991)—or have invaded caves from such subsurface habitats—for example, the amphipod genera *Niphargus* (Karaman and Ruffo 1986) and *Stygobromus* (Holsinger 1986).

Fourth, *Gammarus* species with more southerly distributions are typically in caves and/or springs. In the United States, eleven species occur south of 40°N (roughly the Maryland–Pennsylvania state line). Of these, seven *(G. acherondytes, bousfieldi, desperatus, hyalleloides, minus, pecos,* and *troglophilus)* are associated with caves or springs in areas of exposed carbonate rock (Fig. 5.2). Particularly interesting is the biogeographical pattern of *G. pseudolimnaeus*. In the northern part of its range centered on the Great Lakes, it is found in surface streams. In the southern part of its range in southern Illinois and Missouri, it is typically found in caves and springs. The other two species occurring below 40°N are the southern extensions of two species more widespread in the north—*G. lacustris* in the West and *G. fasciatus* in the East. Neither of these species is associated with karst areas. It is worth recalling that cave and spring habitats are not cooler on average than surface waters in the same area. The difference is that temperatures are buffered in subterranean habitats and approximate the mean annual temperature with relatively little fluctuation (see Chapter 4).

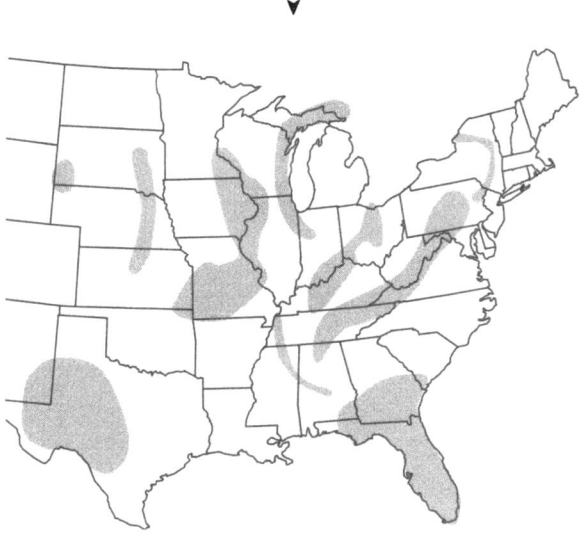

Fig. 5.2. Major exposures of carbonate rock in central and eastern United States. Most of these areas have extensive cave development.

Biogeography of *Gammarus minus*

The range of *G. minus* forms a broad arc from southern Pennsylvania to western Kentucky and southern Indiana, with outlying populations in Missouri to eastern Oklahoma (Fig. 5.3). It is widespread in springs and spring runs, but effectively is limited to carbonate springs with pH greater than 7 (see Chapter 4). There are numerous records of *G. minus* in caves throughout its range. For example, Holsinger, Baroody, and Culver (1976) report *G. minus* from fifty-seven caves in West Virginia. Most of these cave populations show little morphological differentiation from spring populations. Hubricht (1943) suggested that most cave populations had slightly reduced eyes, longer antennae, and a bluish color relative to spring forms. Indeed, Holsinger and Culver (1970) confirmed that eyes are slightly reduced and antennae elongated in most cave populations relative to spring populations. However, it is not clear that all of these differences are genet-

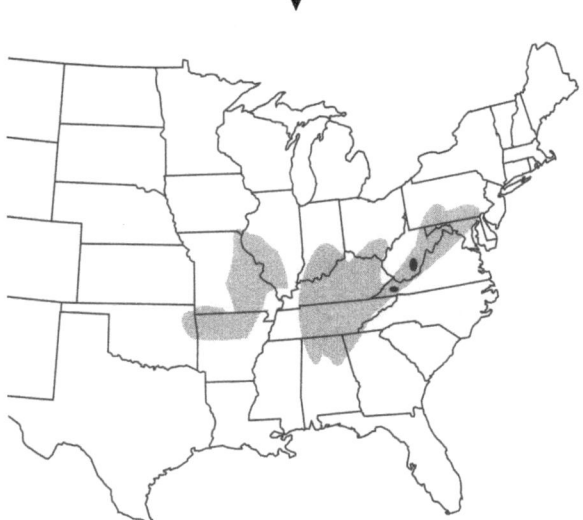

Fig. 5.3. Distribution of *Gammarus minus*. Solid areas within the shading indicate locations of highly modified cave populations, named *tenuipes* by Shoemaker (1940). Modified from Holsinger (1972).

ically determined. Given the profound differences between cave and spring habitats, it would be surprising if there were not environmental effects on morphology. The bluish color of *G. minus* may be due to the absence of the carotenoid precursors required for the carotenoid-protein pigment complexes of amphipods (Beatty 1949). (The genetic determination of morphology is the subject of Chapter 6.)

The cave populations just described are intermediate in morphology between typical spring populations and the highly modified *tenuipes* populations first described by Shoemaker (1940), whereas the highly modified cave populations are limited to two small regions (Fig. 5.3). Why should they be so restricted? Some clues may be obtained from a more careful consideration of these areas. Of the fifty caves in the United States with more than 17 km of known passage (Gulden 1993), nine are within a 30 km radius of Lewisburg, West Virginia, in an area often called Greenbrier Valley karst. This is one of the regions where highly modified *G. minus* populations are found. Here and in other regions with large cave systems there is the subterranean equivalent of a drainage basin (see Ford and Williams 1989). Six of these basins were delineated by W. K. Jones (1973) and are depicted in Figure 5.4. All basins except Rockland Springs have known caves greater than 5 km in length. Culverson Creek basin includes Culverson Creek Cave with more than 30 km of passage (Gulden 1993), and Buckeye Creek basin includes Buckeye Creek Cave, with more than 15 km of passage (Dasher and Balfour 1994). Both of these caves and other caves in the drainage basins have *G. minus* populations, but none is highly modified morphologically. By contrast, the Hole Cave, with more than 35 km of passage, in The Hole basin has a highly modified *G. minus* population. Likewise Benedicts Cave (>20 km), McClung Cave (>25 km), Ludington Cave (>9 km), and Wades Cave (>6 km) in the Davis Spring basin have highly modified populations. Other, smaller caves in the Davis Spring basin, such as Pecks Cave (<0.5 km), have only slightly modified *G. minus* populations, especially on the western side of the basin. Finally, caves in the Davis Hollow basin and the Rockland Springs basin have only slightly modified populations. Thus, highly modified populations of *G. minus* occur only when the following conditions are met:

1. Caves are large, probably at least 2 km in length
2. Basins are relatively large, probably at least 10 km^2
3. Basins have few if any surface streams that flow directly into caves

The difference between Culverson Creek and Buckeye Creek basins on the one hand and The Hole and Davis Spring basins on the other is that the former contains numerous sinking streams.

The other region with highly differentiated cave populations is Ward Cove in Tazewell County, Virginia. In many ways it is the Greenbrier Valley karst in miniature. Maiden Spring drains a karst area of over 25 km^2, with little direct surface input, and includes Fallen Rock Cave with over 10 km of passage (Holsinger 1975).

The three requirements listed above clearly indicate that, in order for troglomorphy in *G. minus* to evolve, there must be isolation of the habitat and the habitat itself must be large. The distribution of *G. minus* provides no support for the traditional notion of the evolution

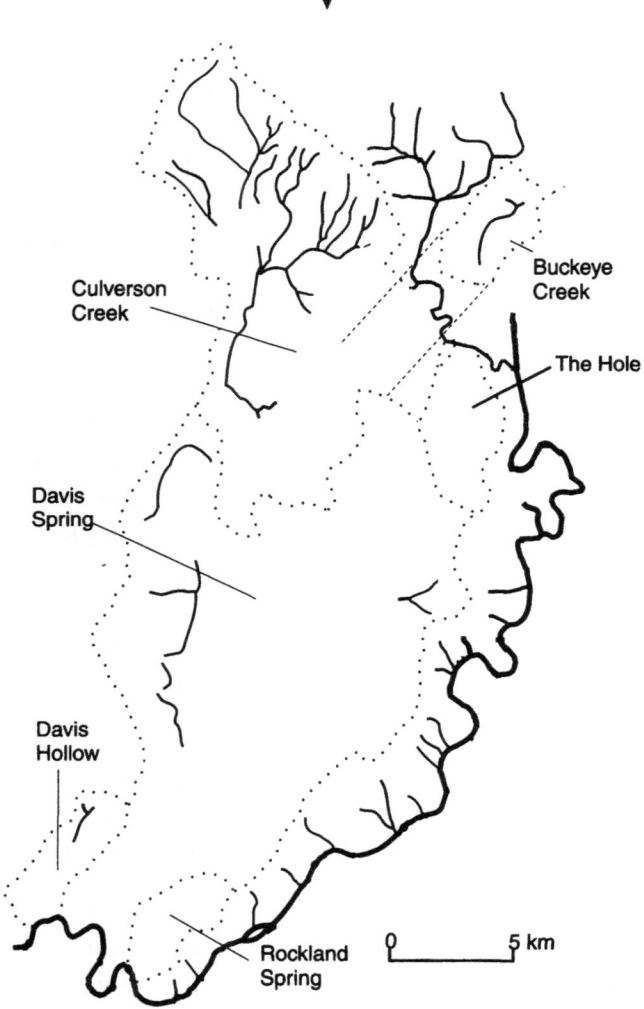

Fig. 5.4. Six subterranean drainage basins in Greenbrier County, West Virginia. Basin boundaries are shown by dotted lines. Surface streams are shown in solid lines—the Greenbrier River is shown at right. Buckeye Creek basin (indicated by dashed lines) lies on top of Culverson Creek. Only The Hole basin and the eastern part of Davis Spring basin contain *Gammarus minus* populations with reduced eyes. Map modified from Jones (1973).

of cave animals—the evolution of small, isolated populations founded by a few individuals resulting in divergence from surface ancestors.

While the three requirements for highly differentiated populations of *G. minus* are necessary for the evolution of troglomorphy, they are clearly not sufficient. In other parts of its range, the species has not invaded large cave systems. Most prominent among these large cave systems is Mammoth Cave. Although *G. minus* is known from springs in the immediate area, it is not found in the cave. It may be that *G. minus* has had fewer opportunities to invade caves in the more southern and western parts of its range. For example, if *G. minus* invaded caves as the result of interglacial warming, it may have been less pronounced in southern and western parts of its range. More likely is the possibility that invasions have been unsuccessful because of competitive and predatory interactions with an old cave stream fauna that includes *Crangonyx* amphipods, *Orconectes* crayfish, and fish in the family Amblyopsidae (Poulson 1992). Culver (1975, 1976) has suggested that many cave stream communities are saturated with species at least in ecological time, making successful invasion difficult.

The highly modified populations of *G. minus* in caves have an extensive distribution within the cave. The best-studied case in this regard is Organ Cave in West Virginia. There are two major stream systems in the cave—the western Lipps drainage and the eastern Organ drainage (Fig. 5.5). For reasons that are not clear, *G. minus* is absent from the Lipps drainage (Fong and Culver 1994). It is nearly ubiquitous in the eastern Organ drainage, although its density varies widely, from $1/m^2$ to well over $100/m^2$. It occurs in at least 10 km of stream and probably twice that. The streams average over 1 m in width, and so, if we use as a conservative estimate a density of $10/m^2$, the total population in Organ Cave is at least 100,000. Compared with populations of invertebrates in surface streams, this may not be remarkable, but it is much larger than most cave populations are thought to be.

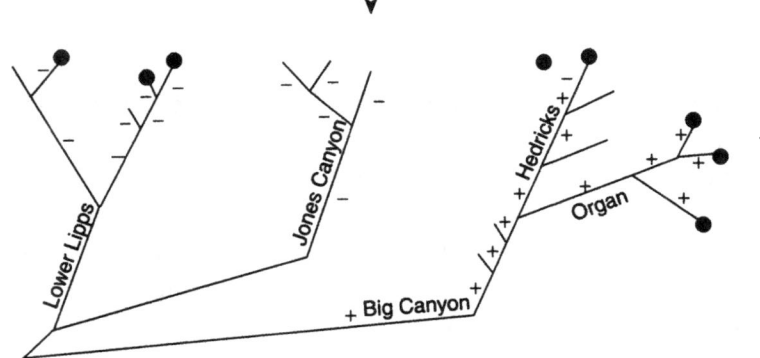

Fig. 5.5. Distribution of *Gammarus minus* (+, presence; −, absence) in Organ Cave. Only the major streams are labeled. Closed circles indicate entrances. Data from Fong and Culver (1994).

Karst Basins

Karst is the name given to landscapes of carbonate rock, such as limestone, that has been molded not by erosion but by chemical solution of bedrock by rain and percolating groundwater. Typical karst basins have been delineated by Jones (1973, Fig. 5.4) in a study of the cave regions developed in thick, flat-bedded limestones in Greenbrier County, West Virginia. Water enters the subterranean drainage largely from vertical percolation of rainfall and runoff through the limestone and to a lesser extent from sinking streams (swallets) at the border of carbonate and noncarbonate rock. It collects in large open channel conduits—that is, cave streams. In the cave systems where *Gammarus minus* is found, the cave streams are typically stony-bottomed streams with alternating riffles (shallow, more turbulent waters over gravel bars) and pools (deeps).

Entrances to caves and cave streams accessible to humans are formed independently of the caves themselves. Curl (1966, 1988) shows that an important fractal property of caves and karst is that most caves have no entrances! Therefore, the actual number of caves and the linear extent of cave streams in a basin is generally much greater than are known. Cave streams generally extend for considerable distance, but they are occasionally interrupted by ceiling collapse, which exposes the cave stream to surface conditions for a short section of its length. These are "karst windows." Eventually, the cave stream plunges below the water table and becomes a closed-channel (water-filled) conduit. The closed-channel conduit ascends and ends at a resurgence (spring).

The factors that result in these cave passage patterns, or "phreatic loops," have been extensively studied by karst geomorphologists (see Ford and Ewers 1978 and Ford and Williams 1989). In brief, the orientation of subsurface water flow is the result of a balance between the direction in which resistance to flow is minimized (maximization of hydraulic conductivity) and the direction in which the rate of energy loss is maximized (the shortest and steepest route) (Ford and Williams 1989, p. 249). The environment of the phreatic loop is quite different from that of open-channel conduits. Velocity is likely higher, since cross-sectional area is constrained, and as a consequence turbulent flow is more likely. White (1988) suggests that flow in karstic tubes greater than 1 cm in diameter is likely to be turbulent. In addition, little sediment is likely present, and in particular the riffle-pool structure of open conduits will be absent. The components of an idealized basin are shown in Figure 5.6.

Most of our work on evolution and adaptation to caves has focused on *G. minus* populations in six subsurface basins in southern West Virginia (Fig. 5.7) and one subsurface basin in Virginia (Fig. 5.8). Typically, water reaches these subsurface basins through percolation, rather than through sinking streams (swallets). These basins range in size from 16 km^2 (II, The Hole basin, Fig. 5.7) to 205 km^2 (III, Davis Spring basin, Fig. 5.7). In all except basin II (The Hole) and basin V (Scott Hollow), water exits at a single resurgence. All resurgences are within 5 m of base level (Greenbrier River, Second Creek, and Spring Creek in West Virginia and Little River in Virginia). All are characterized by a phreatic loop at the downstream end of the basin. Deike (1988) estimates that for basin VI (Organ Cave), the loop is 20 m below the water table at its lowest point.

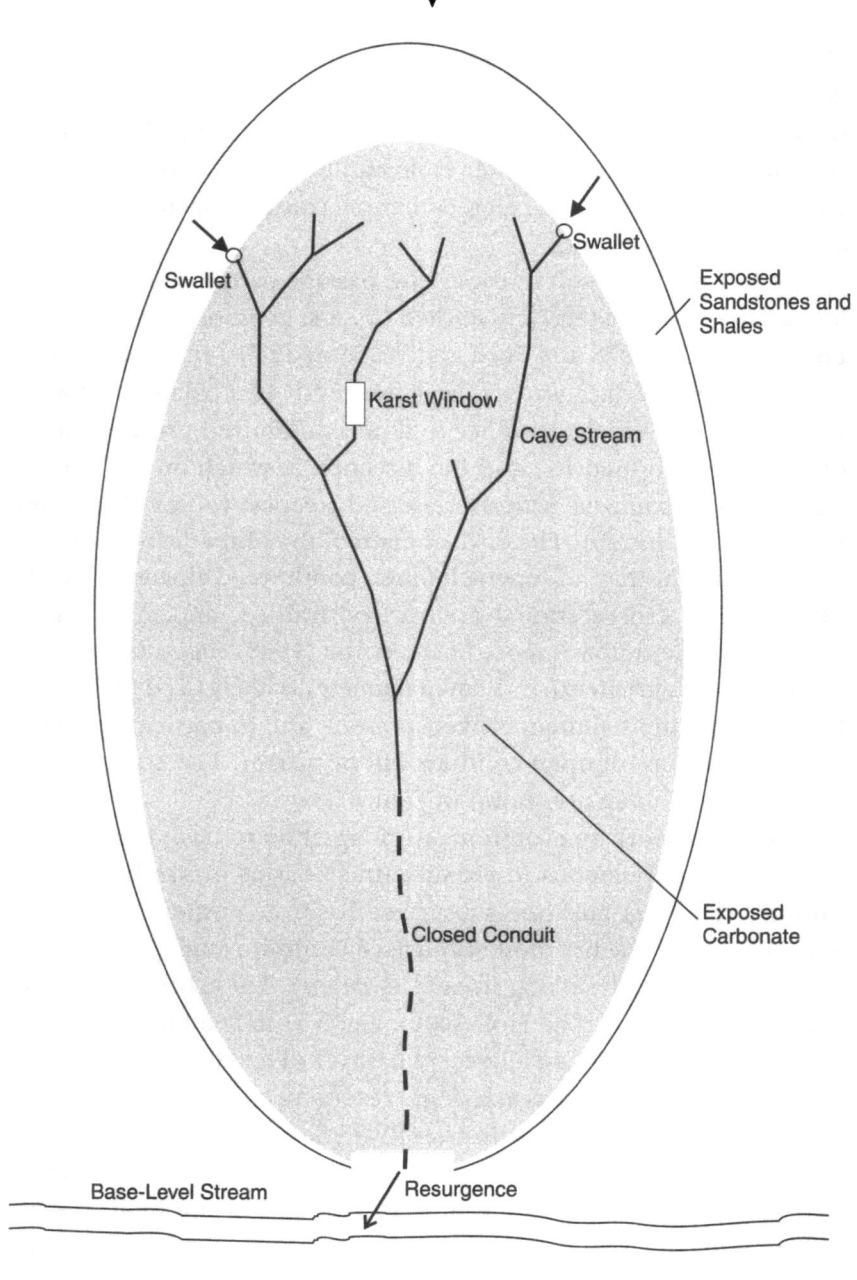

Within a subterranean drainage basin, five distinct habitats can be identified (see Fig. 5.6):

1. Insurgences (swallets)
2. Open conduits (cave streams)
3. Karst windows
4. Closed conduits
5. Resurgences

Extensive sampling of insurgences in Virginia and West Virginia karst areas by Fong has failed to yield any *G. minus* populations. By contrast, open conduits, karst windows, and resurgences typically contain large *G. minus* populations. There are several pieces of evidence indicating that *G. minus* does not regularly inhabit closed conduits (water-filled cave passages). The habitat itself has not been extensively sampled because of its inaccessibility. However, extensive baiting for *G. minus* at the end of the ascending arm of a phreatic loop in the Lost Mill karst area in Tazewell County, Virginia (see Holsinger 1975), adjacent to basin VII (Ward Cove, see Fig. 5.8), yielded less than 10 individuals, rather than the usual hundreds obtained in this way in open conduits. The few individuals obtained were most likely washed out from upstream open-channel habitats. The open-channel habitats in the Lost Mill karst area are inaccessible. Strong indirect evidence includes the rarity of cave forms (less than 1 in 1000) in spring populations. If cave forms occurred in closed conduits contiguous with the resurgence, this frequency should be much higher. The existence of strongly differentiated cave and spring forms makes some physical separation likely.

Fig. 5.6. Idealized karst basin (subterranean area viewed from above). Water entering the basin, indicated by arrows, either flows along impermeable rock (sandstones and shales) and sinks in swallets at points of contact with carbonate rocks *(stippled area)* or percolates vertically into conduits (cave streams). Streams occasionally emerge on the surface at karst windows, where there has been a ceiling collapse in the overhead terrain. After flowing through a closed conduit below the water table, water emerges at a resurgence. Here the resurgence is shown flowing into an aboveground stream.

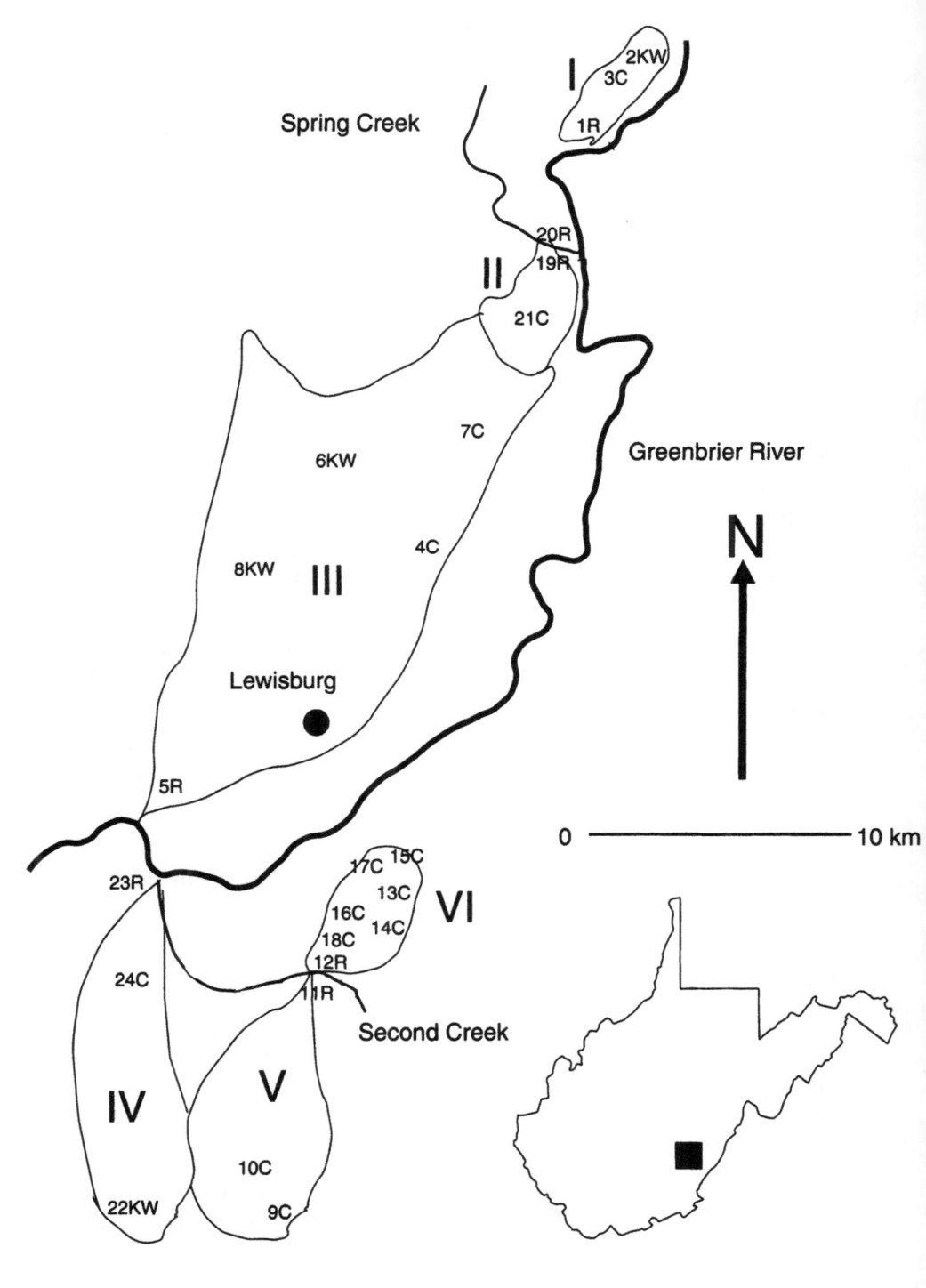

Geographical Genetics of *Gammarus minus*

Our knowledge of genetic differentiation of *Gammarus minus* is largely based on analysis of electrophoretically detectable variation in enzymatic proteins among twenty-four populations in southern West Virginia (Fig. 5.7) and an additional three populations in the Ward Cove karst drainage in Virginia (Fig. 5.8). In assessing genetic structure from these data, we have relied heavily on Wright's (1978)

Fig. 5.7. Six subterranean drainage basins in the study area in southern West Virginia. Site codes are made up of basin number (roman numerals), site number (arabic numerals), and habitat code (C, cave; KW, karst window; R, resurgence). From Kane and Culver (1991).

G. minus localities are as follows:

I1R	Bone-Norman resurgence
I2KW	Taylor Spring karst window
I3C	Bone-Norman Cave
II19R	Burns Cave No. 2 resurgence
II20R	Spring Creek Blue Hole resurgence
II21C	The Hole Cave
III4C	Benedicts Cave
III5R	Davis Spring resurgence
III6KW	Higginbothams Cave karst window
III7C	Ludington Cave
III8KW	Milligan Creek karst window
IV22KW	Boyd Spring karst window
IV23R	Scott Hollow resurgence
IV24C	Scott Hollow Cave
V9C	Hofsackers Cave
V10C	Burnside Branch Cave
V11R	Dickson Spring resurgence
VI12R	Organ Cave resurgence
VI13C	Organ Cave, Organ Stream
VI14C	Organ Cave, 1812 Stream
VI15C	Organ Cave, Sively No. 3 Stream
VI16C	Organ Cave, Hedricks Stream
VI17C	Organ Cave, Masters Stream
VI18C	Organ Cave, Big Canyon

84   The Geography of *Gammarus minus*

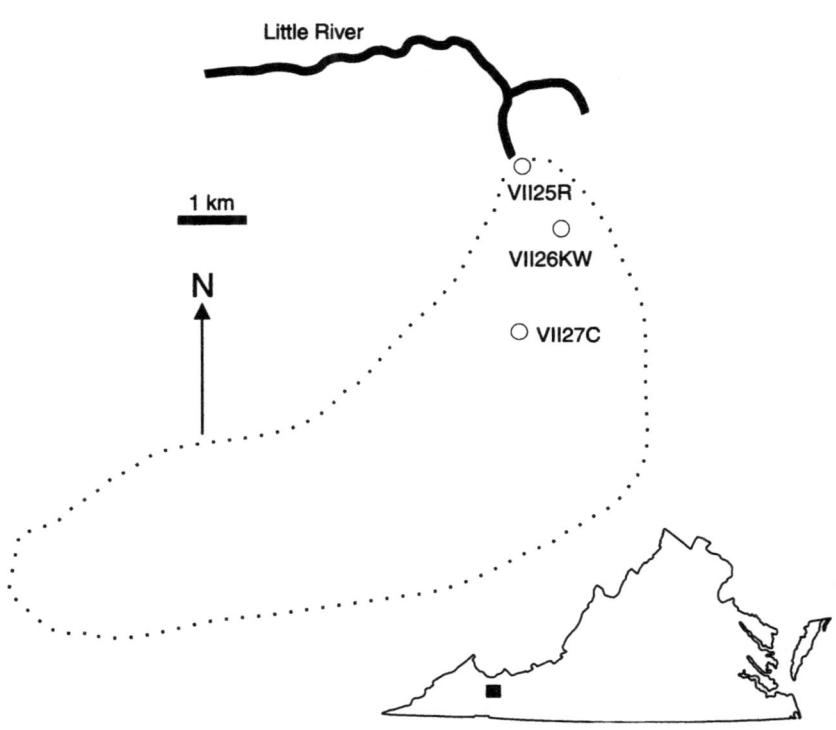

Fig. 5.8. Subterranean drainage basin in southwestern Virginia, near the town of Pounding Mill. Site codes are made up of basin number (roman numerals), site number (arabic numerals), and habitat code (C, cave; KW, karst window; R, resurgence). From Sarbu, Kane, and Culver (1993).

G. minus localities are as follows:
VII25R   Maiden Spring
VII26KW   Hugh Young Cave
VII27C   Fallen Rock Cave

F-statistics. They are a particularly attractive analytical tool because (1) they allow a partitioning of variance and thus allow use of ANOVA techniques (see Cockerum 1969); (2) they equate the between-population component of variance with the relatedness (covariance) of genes within populations; (3) algorithms for computation of F-statistics are readily available (Swofford and Selander 1981, Weir 1990); and (4) they permit simple numerical assessment of the extent of population differentiation and gene flow.

The overall inbreeding coefficient F_{IT} is related to the inbreeding coefficient of each population, F_{IS}, and a measure of population differentiation, F_{ST} (coancestry of individuals within a population), by the following relation:

$$1 - F_{IT} = (1 - F_{IS})(1 - F_{ST}) \qquad (5.1)$$

The within-population inbreeding coefficient F_{IS} is a measure of heterozygote deficiency and is related to effective population size, N_e, by

$$F_{IS} = \frac{1}{2} N_e \qquad (5.2)$$

In Wright's (1978) island model, population differentiation is connected to the average number of migrants per population per generation, Nm, where N is population size and m is migration rate:

$$Nm = (1 - F_{ST})/4F_{ST} \qquad (5.3)$$

Models of genetic differentiation (see Slatkin and Barton 1989) indicate that for $Nm < 0.5$ ($F_{ST} > 0.33$), populations will be unconnected genetically with regard to neutral alleles.

Although F-statistics are a powerful analytical tool, two important assumptions about them must be kept in mind (Weir 1990). First, even though the populations in question may have been distinct for some time, the analysis is built on the assumption that there was a single ancestral population initially. Second, the basic model assumes the absence of selection operating on the loci being analyzed. The difficulty in detecting selection at individual electrophoretic loci has led many—but not all (see Gillespie 1991)—to use neutral models for the analysis of electrophoretic variation.

Using a combination of starch and polyacrylamide gel electrophoresis techniques detailed in Kane, Culver, and Jones (1992), we have resolved 13 enzyme systems encoded by 18 presumptive gene loci in all 27 populations. These include aconitase (ACO, 2 loci); alkaline phophatase (ALP); aldehyde oxidase (AOX, 2 loci); esterase (EST); glutamate dehydrogenase (GDH); glutamate oxaloacetate transaminase (GOT, 2 loci); malate dehydrogenase (MDH); malic enzyme (ME); mannose phosphate isomerase (MPI); peptidase (PEP, 2 loci); 6-phosphogluconate dehydrogenase (PGD); phosphoglucose isomerase (PGI); and superoxide dismutase (SOD, 2 loci). In addition, products of an acid phosphatase (ACP) locus were consistently resolved in the three Ward Cove populations (Sarbu, Kane, and Culver 1993) but could not be consistently resolved in the West Virginia populations. The Virginia and West Virginia studies were done at different times, making it difficult to confirm similarity of electromorphs between the two regions, particularly for highly variable loci. Thus, we present the data sets separately but emphasize the similarities in genetic pattern between the two geographic regions.

The three populations examined in the Ward Cove drainage include one from the resurgence, Maiden Spring, and two upstream subterranean populations: one from Fallen Rock Cave, a large cave with over 8 km of stream passage, and one from Hugh Young Cave, a small cave exiting at a karst window where a large *G. minus* population occurs (Fig. 5.8). Of the nineteen loci examined, ten were polymorphic (0.99 criterion) within or among populations (Table 5.1). A UPGMA dendrogram based on arc distance (Cavalli-Sforza and Edwards 1967) over all nineteen loci yields a pattern of genetic differentiation (Fig. 5.9) that parallels the pattern of morphological differentiation.

Maiden Spring is differentiated from both upstream cave populations (Fig. 5.9), while the two cave populations are genetically quite similar to one another. Overall F_{ST} = 0.226 among the three populations, with between-habitat (spring vs. cave) F_{ST} = 0.204 and within-habitat (Fallen Rock Cave vs. Hugh Young Cave) F_{ST} = 0.022. Overall gene flow is low, with N_m = 0.9, primarily because of the lack of migration between Maiden Spring and the two cave populations. The fixation of alternative alleles at the SOD-2 locus (Table 5.1) between the spring and the cave populations indicates that there may be no gene flow currently between habitats. In contrast, there appears to be a large amount of gene flow between the two cave populations (N_m = 11.1), despite the fact that these populations are farther apart geographically (Fig. 5.8) than are the Hugh Young and Maiden Spring populations.

The total amount of migration between populations is a function of both migration rate (m) and population size (N). Thus, reduced gene flow between cave and spring habitats could result from small population size in either or both habitats. Data on genetic variability in these populations suggest that this is unlikely, however. Values for the average proportion of polymorphic loci per population (P) are similar among Fallen Rock (0.47) and Hugh Young (0.42) caves and Maiden Spring (0.47). The average proportion of heterozygous loci per individual (H) and the average number of alleles per locus (n_A) did not differ significantly (by the Kruskal-Wallis test) among Fallen Rock Cave (H = 0.103, n_A = 2.0), Hugh Young Cave (H = 0.102, n_A = 1.9), and Maiden Spring (H = 0.075, n_A = 2.1). Not only are these variability values similar among all three populations, but they also fall in the range of values commonly reported for widely distributed and abundant populations of surface-dwelling invertebrates (Nevo, Beiles, and Ben-Shlomo 1984).

These data indicate that the Ward Cove *G. minus* populations are not excessively small, nor do they appear to have undergone recent bottlenecks in population size. Assuming genetic equilibrium in these populations and selective neutrality of the electrophoretic variation, average heterozygosity (H) can be related to effective population size (N_e) and the mutation rate (v) to electrophoretically detectable neutral alleles by:

Table 5.1. Electromorph frequencies for three populations of *G. minus* in Ward Cove, Virginia. Alleles indicated by letters A, B, C, etc. N = sample size; see Fig. 5.8 for locations.

Gene locus	Population		
	Maiden Spring (VII25R)	Hugh Young Cave karst window (VII26KW)	Fallen Rock Cave (VII27C)
ACO-1			
N	68	89	58
A	0.007	0.000	0.009
B	0.868	0.775	0.586
C	0.118	0.219	0.405
D	0.007	0.006	0.000
ACO-2			
N	65	80	57
A	0.000	0.000	0.009
B	0.038	0.063	0.018
C	0.946	0.931	0.956
D	0.015	0.006	0.018
ACP			
N	64	61	61
A	0.078	0.139	0.164
B	0.922	0.861	0.836
GOT-1			
N	48	63	51
A	0.000	0.024	0.000
B	0.042	0.079	0.108
C	0.938	0.849	0.853
D	0.020	0.048	0.039
GOT-2			
N	48	60	50
A	0.052	0.008	0.000
B	0.938	0.617	0.520
C	0.010	0.367	0.420
D	0.000	0.000	0.060
E	0.000	0.008	0.000

Table 5.1 (continued).

	Population		
Gene locus	Maiden Spring (VII25R)	Hugh Young Cave karst window (VII26KW)	Fallen Rock Cave (VII27C)
MDH			
N	55	71	58
A	0.018	0.014	0.017
B	0.973	0.873	0.819
C	0.009	0.028	0.034
D	0.000	0.085	0.129
MPI			
N	44	52	48
A	0.034	0.000	0.000
B	0.875	0.846	0.927
C	0.091	0.154	0.073
PEP-1			
N	74	99	64
A	0.000	0.000	0.008
B	0.919	0.955	0.898
C	0.020	0.000	0.047
D	0.061	0.045	0.047
PGI			
N	61	56	58
A	0.025	0.009	0.009
B	0.033	0.000	0.000
C	0.664	0.991	0.983
D	0.066	0.000	0.009
E	0.213	0.000	0.000
SOD-2			
N	66	74	67
A	0.000	1.000	1.000
B	1.000	0.000	0.000

90 The Geography of *Gammarus minus*

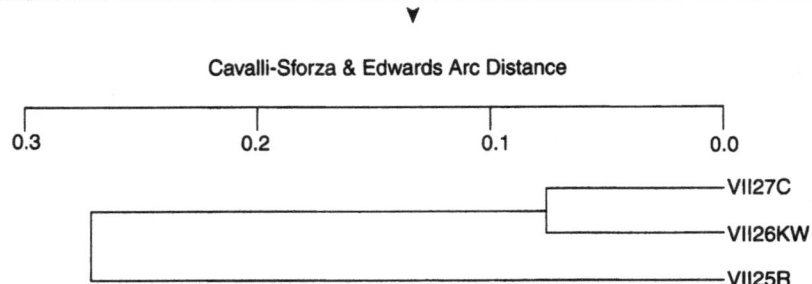

Fig. 5.9. Dendrogram of Cavalli-Sforza and Edwards arc distances for electrophoretic data (Table 5.1) for Ward Cove *Gammarus minus* populations (Fig. 5.8). From Sarbu, Kane, and Culver (1993).

$$H = (4N_e\nu)/(4N_e\nu + 1) \tag{5.4}$$

Assuming $\nu = 1 \times 10^{-7}$ (Kimura 1983), we estimate the effective population size of the Ward Cove populations to lie in the range of 1.1×10^5 and 1.7×10^5.

Genetic patterns among the twenty-four West Virginia populations (Tables 5.2A–D) resemble the pattern observed in the Ward Cove drainage. The UPGMA dendrogram of the electrophoretic data (Fig. 5.10) show that resurgence populations, with one exception, are differentiated from upstream cave and karst-window populations. Upstream cave and karst-window populations within a drainage tend to cluster together genetically (Fig. 5.10). The exceptional cases are also informative. The cave population and the karst-window population of basin I cluster with the resurgence populations genetically. Interestingly, if the number of ommatidia is used as an index (see below), these populations are undifferentiated morphologically from the resurgence population (I1R) as well. The only basin within which there is differentiation among cave and karst-window populations is basin III (Fig. 5.10), which is also the largest (Fig. 5.7) and hydrologically most complex (see Heller 1991) of the six basins. Differentiation between the two southernmost populations in the drainage (III4C and II8KW, Fig. 5.7) reflects the presence of two parallel subterranean flows (east and west sides of the basin) that converge at the resurgence. The northern populations (III6KW and III7C, Fig. 5.7)

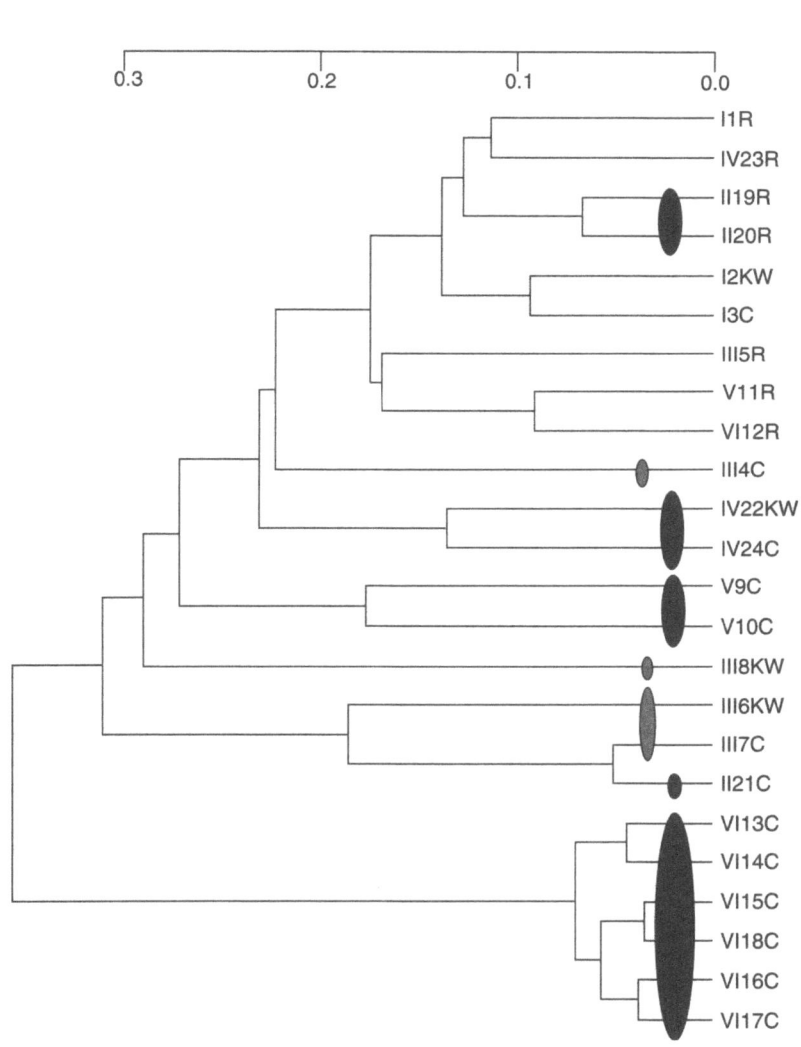

Fig. 5.10. Dendrogram of Cavalli-Sforza and Edwards arc distances for electrophoretic data for West Virginia *Gammarus minus* listed in Fig. 5.7. Solid ellipses cover populations in individual basins; hatched ellipses cover populations in basin III (Fig. 5.7). From Kane and Culver (1991).

Table 5.2A. Electromorph frequencies for six West Virginia populations of *G. minus*. Alleles indicated by letters A, B, C, etc. Population codes are given in Fig. 5.7. *N* = sample size.

Gene locus	Population					
	I1R	I2KW	I3C	III4C	III5R	III6KW
ACO-1						
N	60	53	39	63	85	49
A	.000	.113	.051	.000	.071	.000
B	.833	.764	.538	.008	.894	.561
C	.158	.123	.410	.944	.035	.429
D	.008	.000	.000	.048	.000	.010
ACO-2						
N	49	50	40	54	43	48
A	.000	.000	.000	.000	.000	.000
B	.000	.000	.000	.000	.000	.021
C	.980	1.000	1.000	1.000	1.000	.979
D	.020	.000	.000	.000	.000	.000
E	.000	.000	.000	.000	.000	.000
GOT-1						
N	42	62	40	55	54	46
A	.036	.016	.000	.000	.204	.000
B	.000	.000	.000	.000	.000	.000
C	.964	.968	1.000	1.000	.778	1.000
D	.000	.016	.000	.000	.019	.000
GOT-2						
N	39	44	39	52	42	44
A	.000	.000	.000	.000	.000	.000
B	.000	.045	.115	.000	.000	.000
C	.000	.000	.000	.000	.000	.000
D	.987	.955	.872	1.000	.917	1.000
E	.013	.000	.013	.000	.071	.000
F	.000	.000	.000	.000	.012	.000
MDH-1						
N	51	67	40	58	61	46
A	.000	.000	.000	.000	.000	.000
B	.000	.000	.013	.009	.008	.000
C	1.000	1.000	.988	.991	.992	1.000
D	.000	.000	.000	.000	.000	.000
E	.000	.000	.000	.000	.000	.000

Table 5.2A (continued).

Gene locus	Population					
	I1R	I2KW	I3C	III4C	III5R	III6KW
MPI-1						
N	40	56	52	40	64	44
A	.000	.009	.000	.000	.000	.000
B	.013	.018	.000	.000	.023	.034
C	.988	.938	.769	1.000	.945	.966
D	.000	.036	.231	.000	.031	.000
PEP-1						
N	41	40	45	45	44	50
A	.000	.000	.000	.000	.159	.000
B	.646	.237	.200	1.000	.795	.990
C	.000	.000	.000	.000	.000	.000
D	.354	.762	.789	.000	.045	.010
E	.000	.000	.011	.000	.000	.000
PGD-1						
N	40	64	45	38	41	47
A	.000	.000	.000	.000	.000	.000
B	1.000	1.000	1.000	1.000	1.000	1.000
PGI-1						
N	47	61	44	42	73	41
A	.000	.000	.000	.000	.014	.000
B	.032	.016	.011	.083	.377	.000
C	.000	.041	.000	.000	.000	.000
D	.138	.082	.114	.000	.000	.000
E	.798	.828	.830	.881	.541	1.000
F	.032	.016	.034	.024	.068	.000
G	.000	.008	.011	.012	.000	.000
H	.000	.000	.000	.000	.000	.000
I	.000	.008	.000	.000	.000	.000
SOD-1						
N	52	36	40	39	20	58
A	1.000	1.000	1.000	1.000	1.000	.000
B	.000	.000	.000	.000	.000	1.000

Table 5.2B. Electromorph frequencies for six West Virginia populations of *G. minus*. Alleles indicated by letters A, B, C, etc. Population codes are given in Fig. 5.7. *N* = sample size.

Gene locus	Population					
	III7C	III8KW	V9C	V10C	V11R	VI12R
ACO-1						
N	39	67	38	50	62	57
A	.000	.082	.000	.000	.032	.000
B	.808	.627	.697	.690	.661	.816
C	.192	.276	.303	.310	.298	.184
D	.000	.015	.000	.000	.008	.000
ACO-2						
N	38	50	42	51	50	45
A	.000	.000	.000	.000	.000	.000
B	.000	.050	.000	.000	.020	.056
C	1.000	.930	.333	.951	.750	.767
D	.000	.010	.000	.020	.050	.000
E	.000	.010	.667	.029	.180	.178
GOT-1						
N	40	53	40	44	70	83
A	.000	.057	.000	.000	.036	.042
B	.000	.000	.138	.000	.000	.000
C	1.000	.925	.863	1.000	.964	.958
D	.000	.019	.000	.000	.000	.000
GOT-2						
N	40	45	40	42	44	57
A	.000	.011	.000	.012	.023	.000
B	.000	.022	.000	.000	.000	.053
C	.000	.000	.000	.000	.000	.009
D	1.000	.956	1.000	.988	.864	.904
E	.000	.011	.000	.000	.091	.035
F	.000	.000	.000	.000	.023	.000
MDH-1						
N	45	73	42	56	50	58
A	.000	.000	.000	.000	.000	.009
B	.000	.151	.000	.000	.020	.017
C	1.000	.048	.762	.830	.970	.974
D	.000	.795	.000	.000	.000	.000
E	.000	.007	.238	.170	.010	.000

Table 5.2B (continued).

Gene locus	Population					
	III8C	III8KW	V9C	V10C	V11R	VI12R
MPI-1						
N	40	43	40	43	48	42
A	.000	.000	.000	.000	.000	.000
B	.000	.105	.000	.465	.000	.036
C	1.000	.744	1.000	.535	1.000	.964
D	.000	.151	.000	.000	.000	.000
PEP-1						
N	45	49	45	39	40	38
A	.000	.020	.000	.000	.000	.013
B	1.000	.755	1.000	1.000	1.000	.974
C	.000	.214	.000	.000	.000	.000
D	.000	.010	.000	.000	.000	.013
E	.000	.000	.000	.000	.000	.000
PGD-1						
N	40	50	40	39	48	61
A	.000	.080	.000	.000	.000	.000
B	1.000	.920	1.000	1.000	1.000	1.000
PGI-1						
N	44	54	40	43	45	53
A	.000	.019	.000	.000	.000	.000
B	.000	.167	.000	.000	.000	.009
C	.000	.000	.000	.000	.000	.000
D	.000	.306	.000	.000	.022	.038
E	.068	.491	.000	.000	.700	.660
F	.932	.019	1.000	1.000	.267	.236
G	.000	.000	.000	.000	.011	.038
H	.000	.000	.000	.000	.000	.019
I	.000	.000	.000	.000	.000	.000
SOD-1						
N	45	38	40	49	40	35
A	.000	1.000	1.000	1.000	1.000	1.000
B	1.000	.000	.000	.000	.000	.000

Table 5.2C. Electromorph frequencies for six West Virginia populations of *G. minus*. Alleles indicated by letters A, B, C, etc. Population codes are given in Fig. 5.7. *N* = sample size.

Gene locus	Population					
	VI13C	VI14C	VI15C	VI16C	VI17C	VI18C
ACO-1						
N	45	43	50	80	48	40
A	.000	.000	.000	.006	.000	.000
B	.567	.721	.450	.813	.750	.563
C	.433	.279	.550	.181	.250	.438
D	.000	.000	.000	.000	.000	.000
ACO-2						
N	53	50	58	49	58	40
A	.000	.000	.000	.000	.000	.000
B	.000	.000	.000	.000	.000	.000
C	.000	.000	.000	.000	.000	.000
D	.009	.000	.000	.000	.000	.000
E	.991	1.000	1.000	1.000	1.000	1.000
GOT-1						
N	50	52	45	67	53	40
A	.180	.279	.300	.254	.245	.325
B	.000	.000	.000	.000	.000	.000
C	.820	.721	.700	.746	.755	.675
D	.000	.000	.000	.000	.000	.000
GOT-2						
N	40	40	39	27	41	37
A	.000	.000	.000	.000	.000	.000
B	.000	.000	.000	.000	.000	.000
C	.000	.000	.000	.000	.000	.000
D	.325	.275	.333	.222	.207	.338
E	.675	.725	.667	.778	.793	.662
F	.000	.000	.000	.000	.000	.000
MDH-1						
N	51	56	45	66	40	40
A	.000	.000	.000	.000	.000	.000
B	.000	.000	.000	.000	.000	.000
C	1.000	1.000	1.000	1.000	1.000	1.000
D	.000	.000	.000	.000	.000	.000
E	.000	.000	.000	.000	.000	.000

The Geography of *Gammarus minus* 97

Table 5.2C (continued).

Gene locus	Population					
	VI13C	VI14C	VI15C	VI16C	VI17C	VI18C
MPI-1						
N	52	43	50	45	40	40
A	.000	.000	.000	.000	.000	.000
B	.000	.000	.000	.011	.000	.000
C	1.000	1.000	1.000	.989	1.000	1.000
D	.000	.000	.000	.000	.000	.000
PEP-1						
N	45	44	52	47	55	43
A	.000	.000	.000	.000	.000	.000
B	.133	.239	.192	.255	.173	.209
C	.000	.000	.000	.000	.000	.000
D	.867	.761	.808	.745	.827	.791
E	.000	.000	.000	.000	.000	.000
PGD-1						
N	56	35	40	49	40	45
A	.000	.000	.000	.000	.000	.000
B	1.000	1.000	1.000	1.000	1.000	1.000
PGI-1						
N	41	41	42	55	43	44
A	.000	.000	.000	.000	.000	.000
B	.000	.000	.000	.000	.000	.000
C	.000	.000	.000	.000	.000	.000
D	.000	.000	.000	.000	.000	.000
E	.000	.012	.071	.127	.221	.193
F	.512	.463	.571	.645	.488	.557
G	.488	.524	.357	.227	.291	.250
H	.000	.000	.000	.000	.000	.000
I	.000	.000	.000	.000	.000	.000
SOD-1						
N	45	45	73	35	58	40
A	1.000	1.000	1.000	1.000	1.000	1.000
B	.000	.000	.000	.000	.000	.000

Table 5.2D. Electromorph frequencies for six West Virginia populations of *G. minus*. Alleles indicated by letters A, B, C, etc. Population codes are given in Fig. 5.7. *N* = sample size.

Gene locus	Population					
	II19R	II20R	II21C	IV22KW	IV23R	IV24C
ACO-1						
N	44	42	40	56	63	46
A	.080	.083	.000	.196	.000	.000
B	.716	.750	.712	.777	.929	.957
C	.205	.167	.287	.027	.071	.043
D	.000	.000	.000	.000	.000	.000
ACO-2						
N	37	39	40	35	58	43
A	.000	.013	.000	.000	.000	.012
B	.095	.103	.000	.000	.000	.000
C	.905	.885	1.000	1.000	.974	.988
D	.000	.000	.000	.000	.000	.000
E	.000	.000	.000	.000	.026	.000
GOT-1						
N	43	40	38	52	44	49
A	.047	.038	.000	.106	.284	.020
B	.000	.000	.000	.000	.000	.000
C	.953	.962	1.000	.894	.705	.980
D	.000	.000	.000	.000	.011	.000
GOT-2						
N	41	39	38	40	42	40
A	.000	.000	.000	.000	.048	.000
B	.000	.000	.000	.013	.036	.000
C	.000	.000	.000	.000	.000	.000
D	.768	.821	1.000	.788	.869	.363
E	.024	.000	.000	.200	.012	.637
F	.207	.179	.000	.000	.036	.000
MDH-1						
N	50	48	41	64	48	40
A	.000	.000	.000	.000	.000	.000
B	.000	.000	.000	.023	.000	.000
C	1.000	1.000	1.000	.977	1.000	1.000
D	.000	.000	.000	.000	.000	.000
E	.000	.000	.000	.000	.000	.000

Table 5.2D (continued).

Gene locus	Population					
	II19R	II20R	II21C	IV22KW	IV23R	IV24C
MPI-1						
N	44	42	40	50	43	45
A	.000	.000	.000	.000	.000	.000
B	.000	.012	.000	.000	.000	.000
C	.977	.976	1.000	1.000	.988	1.000
D	.023	.012	.000	.000	.012	.000
PEP-1						
N	41	41	39	30	45	43
A	.000	.024	.013	.000	.011	.000
B	.549	.512	.987	1.000	.667	1.000
C	.000	.000	.000	.000	.000	.000
D	.451	.463	.000	.000	.322	.000
E	.000	.000	.000	.000	.000	.000
PGD-1						
N	42	40	39	36	45	44
A	.000	.000	.000	.000	.000	.000
B	1.000	1.000	1.000	1.000	1.000	1.000
PGI-1						
N	42	47	43	57	56	49
A	.000	.000	.000	.000	.000	.000
B	.012	.011	.000	.009	.000	.000
C	.000	.000	.000	.000	.071	.000
D	.000	.021	.000	.026	.116	.000
E	.976	.809	.244	.035	.625	.214
F	.012	.160	.733	.737	.188	.276
G	.000	.000	.023	.193	.000	.510
H	.000	.000	.000	.000	.000	.000
I	.000	.000	.000	.000	.000	.000
SOD-1						
N	40	36	49	30	43	50
A	1.000	1.000	.000	1.000	1.000	1.000
B	.000	.000	1.000	.000	.000	.000

actually cluster with the cave population of basin II (Fig. 5.10). Subterranean drainage evolution has been such that the drainage divide between basins II and III has been moving northward. Thus northern populations of basin III have probably been captured, hydrologically, from basin II (Heller 1991).

A significant advantage of the West Virginia populations is their division among distinct hydrological basins. It is possible, using hierarchical F-statistics (Wright 1978), to assess the effect of hydrologically mediated gene flow (migration among hydrologically connected populations irrespective of habitat) and the effect of habitat differences (differentiation among populations in different habitats in spite of any hydrological connection) on the observed pattern of genetic differentiation. Cave and karst-window populations within a basin are potentially connected hydrologically, but they are hydrologically isolated from cave and karst-window populations of other basins. Thus sets of cave and karst-window populations within basins produce six distinct hydrologic entities (Fig. 5.7). The downstream resurgences of different basins are connected to one another hydro-

Table 5.3A. Hierarchical F-statistic (Wright 1978) analyses of electrophoretic data for twenty-four populations of *G. minus* in West Virginia (see Fig. 5.7): Analysis of differentiation of populations within and among seven drainage units (six basins plus resurgences).

Locus	$F_{pop\text{-}basin}$	$F_{basin\text{-}tot}$	$F_{pop\text{-}tot}$
ACO-1	0.142	0.035	0.163
ACO-2	0.260	0.743	0.810
GOT-1	0.052	0.083	0.130
GOT-2	0.061	0.419	0.454
MDH-1	0.576	−0.016	0.569
MPI-1	0.195	0.045	0.231
PEP-1	0.128	0.453	0.523
PGD-1	0.072	−0.008	0.065
PGI-1	0.220	0.256	0.419
SOD-1	1.000	0.462	1.000
Total	0.255	0.330	0.480

logically by the network of surface streams. However, they are isolated from upstream cave and karst-window populations by inhospitable closed conduits. Thus the seven resurgence populations collectively form a seventh distinct hydrological entity. Alternatively, populations can be distinguished according to habitat type. On this basis seven populations can be classified as occurring in resurgences, four in karst window habitats, and the remaining thirteen in caves.

When F-statistics are calculated hierarchically on the basis of hydrology (Table 5.3A), there is considerable differentiation among populations within a hydrological unit (among resurgence populations and among cave and karst-window populations within a basin, $F_{pop\text{-}basin}$), and an even larger component of differentiation among hydrologic units ($F_{basin\text{-}tot}$). Of the total among-population genetic variance, approximately 31 percent occurs within hydrologic units and 69 percent occurs among hydrologic units. Much of the within–hydrologic unit variance is a consequence of differentiation among populations of basin III, which, as we indicated previously, appears to have a hydrological explanation itself. Thus a large amount of the

Table 5.3B. Hierarchical F-statistic (Wright 1978) analyses of electrophoretic data for twenty-four populations of *G. minus* in West Virginia (see Fig. 5.7): Analysis of differentiation of populations within and among three habitat types (resurgences, karst windows, and caves).

Locus	$F_{pop\text{-}hab}$	$F_{hab\text{-}tot}$	$F_{pop\text{-}tot}$
ACO-1	0.152	0.012	0.163
ACO-2	0.774	0.159	0.810
GOT-1	0.139	−0.010	0.130
GOT-2	0.412	0.072	0.454
MDH-1	0.550	0.042	0.569
MPI-1	0.251	−0.026	0.231
PEP-1	0.537	−0.029	0.523
PGD-1	0.058	0.008	0.065
PGI-1	0.339	0.121	0.419
SOD-1	1.000	−0.063	1.000
Total	0.452	0.051	0.480

genetic differentiation among populations is congruent with present and historical hydrologic patterns that presumably reflect present and past opportunities for gene flow.

If the differentiation in enzymatic proteins as determined by electrophoresis was primarily due to differences in the selection pressures in resurgences, karst windows, and caves, we would anticipate that a large component of the genetic variance would be apportioned among habitats. However, the hierarchical F-statistic analysis based on habitat yields a large within-habitat component ($F_{pop\text{-}hab}$) and a much smaller among-habitat component ($F_{hab\text{-}tot}$, Table 5.3B). Overall, approximately 89 percent of the among-population genetic variance is partitioned among populations within habitat type, and only 11 percent is partitioned among habitats. This is not to say that natural selection on allozyme loci can be ruled out completely. Some loci (for example, ACO-2 and PGI, Table 5.3B) do, in fact show relatively large values of F for the among-habitat component. In these cases, however, there is an even larger among-population within-habitat component (80 percent of the total among-population variance for ACO-2 and 71 percent for PGI). Thus, the overriding influence on the genetic structure of these populations appears to be gene flow between hydrologically connected populations of *G. minus* rather than different selective regimes in different habitats.

Genetic variability levels in the West Virginia populations are relatively large in general (Fig. 5.11), and comparable in magnitude to those in the Ward Cove populations. There are no differences in heterozygosity among caves (median = 0.087), resurgences (median = 0.086) and karst windows (median = 0.65). Despite this general finding of relatively high heterozygosity levels, four populations exhibit reduced levels of genetic variability (H = 0.020, range = 0.008–0.028). These include the cave population sampled in basin II (II21C, Fig. 5.7) and the three northernmost populations of basin III (III4C, III6KW, III7C of Fig. 5.7). The four populations occur in the region where hydrologic evolution appears to be ongoing, and in fact three of the populations cluster together genetically (all except III4C in Fig. 5.10), despite being in two different basins at the present time.

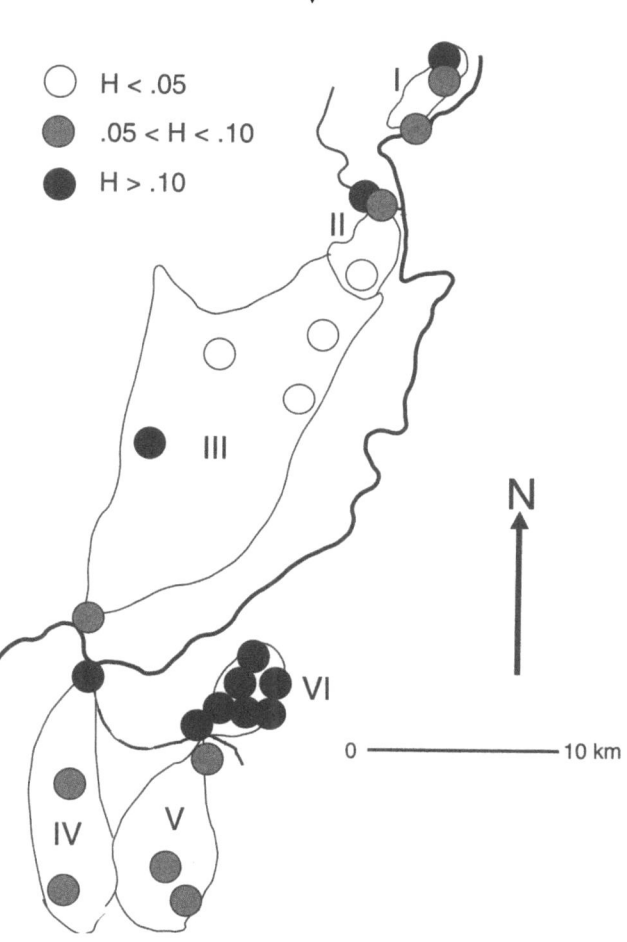

Fig. 5.11. Expected (Hardy-Weinberg) heterozygosity in twenty-four *G. minus* populations (see Fig. 5.7).

We presume that this reflects colonization of these habitats at a very early point in basin evolution, suggesting that these populations are the consequence of one of the oldest subterranean invasions. Further support for this contention is the fact that the cave populations in question show extreme eye reduction, with less than 2 ommatidia per eye (see below). Continuing low levels of variability in these populations suggest that they have little gene flow with other populations in the drainages, as well as little gene flow with each other. In fact, the present hydrologic relationships indicate that one of these populations (II21C) is now completely isolated from the other three.

In sum, the genetic patterns among *G. minus* populations, as revealed by the electrophoretic studies, are largely congruent with past and present opportunities for gene flow. Differentiation of resurgence populations from upstream cave and karst-window populations in a basin, in both Virginia and West Virginia, reflects the change from closed- to open-conduit systems as one proceeds upstream, and the inhospitable nature of the former habitat type to *G. minus* populations. In West Virginia, where multiple basins were studied, partitioning of genetic variability is by hydrologic unit rather than by habitat type, indicating that opportunity for gene flow, as opposed to differences in selection pressures, is the primary factor producing the observed genetic structure. In only one instance (populations of basins II and III) was a genetic result in conflict with present hydrology, but historical changes in drainage pattern, producing historical changes in gene flow, are sufficient to explain this anomaly. Although we cannot completely rule out some effect of selection on allozyme variability, it appears that such an effect would be minor compared with the effect of hydrologically mediated gene flow.

Geographic Morphology of *Gammarus minus*

Eye and size characters have been surveyed for many of the West Virginia populations shown in Figure 5.7. In the most extensive study to date, Culver (1987) summarized variation in head length, eye area, and number of ommatidia (the individual units in a compound eye) in four cave and four spring populations from four basins (II, III, V, and VI in Fig. 5.7). It should come as no surprise that for each basin studied, there were significant differences in the relation between head length and both eye area and ommatidia number. A typical example is shown in Figure 5.12 for the relation between eye area and

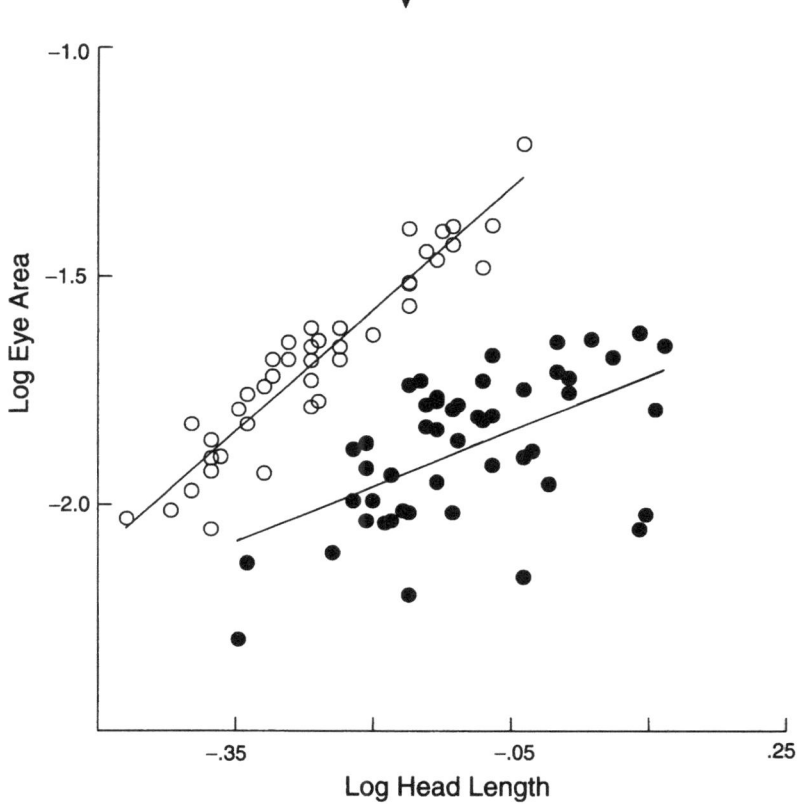

Fig. 5.12. Relationship between head length and eye area (both in log scale) for *G. minus* from Organ Cave *(solid circles)* and Organ Cave resurgence *(open circles)*. Lines are least-squares fit for log y = log a + b log x.

head length for the Organ Cave population (VI13C in Fig. 5.7) and the Organ Cave resurgence (VI12R in Fig. 5.7). For all resurgence populations, eyes were larger and had more ommatidia.

Differences among Cave Populations

There were significant differences among cave populations as well. For the linear relationship between head length *(x)* and eye area *(y)* with the two fitted parameters *a* and *b*,

$$y = ax^b \qquad (5.5)$$

the allometric coefficients (b's) were significantly different among the four cave populations (Fig. 5.13), ranging from 0.49 (negative allometry) in Benedict's Cave to 1.33 (positive allometry) in The Hole. The allometric coefficients for cave populations were less than

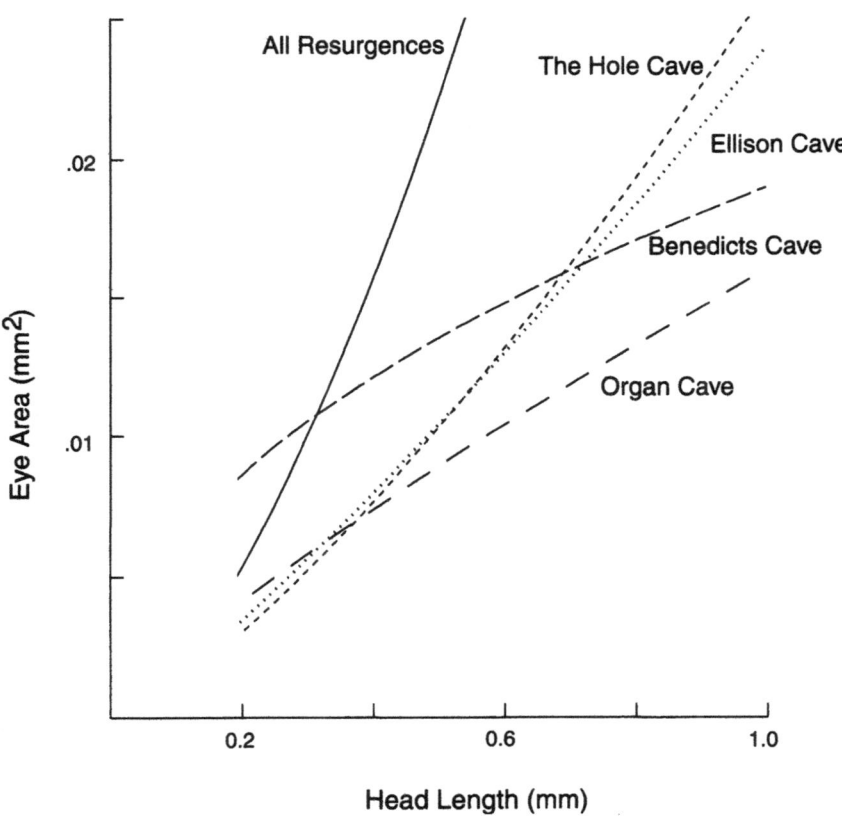

Fig. 5.13. Allometric curves ($y = ax^b$) of eye area and head length for four cave populations of *Gammarus minus* from four basins (II, III, V, and VI of Fig. 5.7). Allometric curves for the four resurgence populations are not significantly different one from the other. Data from Culver (1987).

that for the corresponding resurgence population in three of four cases. In general there is no sign of convergence of the coefficients a and b in equation (5.5), although all cave populations had reduced eye area (y).

Even more striking were differences between cave populations in ommatidia structure and numbers. The number of distinguishable pigment bundles (retinular cells) was counted (described as "eye facets") for the same four cave populations (Culver 1987). For two populations (Ellison Cave and Benedicts Cave), the allometry coefficient was not different from zero. In one population (Organ Cave), the number of retinular cells actually declined with head size (negative b), and in only one population (The Hole Cave) was the allometry coefficient non-negative. Histograms of retinular cell numbers for the four cave populations show the differences very clearly (Fig. 5.14). Two populations (Ellison Cave and The Hole Cave) have unimodal distributions with few individuals lacking retinular cells in either population. In general, Ellison Cave individuals had more pigment bundles. This suggests that there were differences in the time of invasion of these two basins, all else being equal. Individuals from Organ Cave also had a unimodal distribution of pigment cell number, but with over 15 percent of individuals lacking discernible pigment bundles. In contrast, Benedicts Cave had a bimodal distribution. These differences among cave populations of different basins suggest independent invasions that have resulted in roughly the same overall level of eye reduction but by different mutational pathways.

Information on internal anatomy of the eye supports the hypothesis of separate invasions as well. Of the four populations examined for external anatomy, all but Ellison Cave were also studied with respect to internal eye structure, as was the population from Fallen Rock Cave in Virginia (see Fig. 5.8). Once again, differences in overall levels of degeneration as well as in details of the degeneration are apparent (Table 5.4). Internally, individuals from Organ Cave had lost more optic elements while the other three populations differed primarily in the degree of degeneration of various elements.

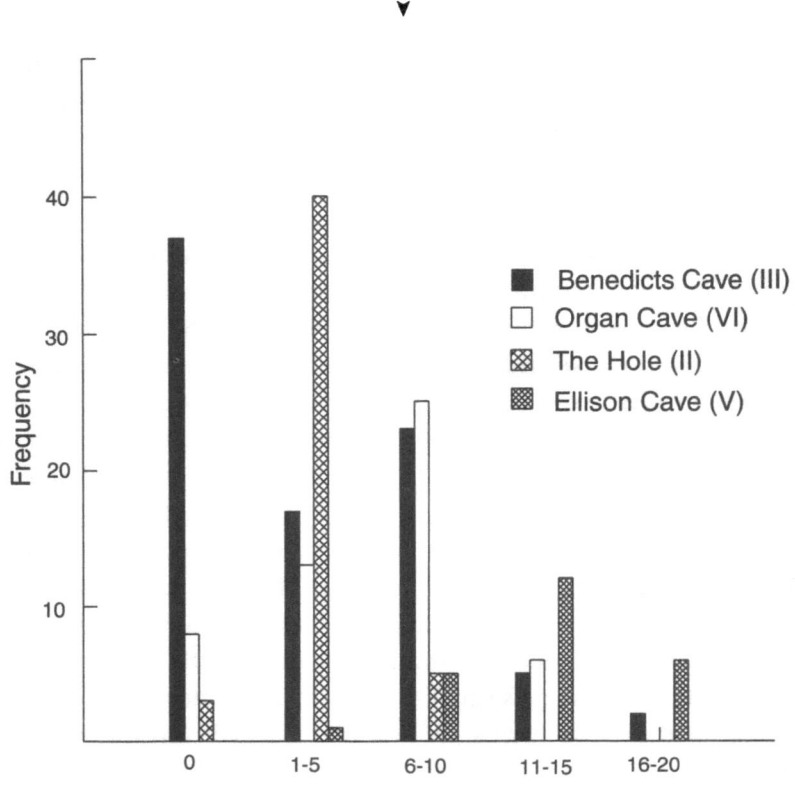

Fig. 5.14. Histogram of eye pigment bundle number (retinular cells) in four populations of *G. minus.* Roman numerals refer to basin (see Fig. 5.7). Ellison Cave is a cave near Burnside Branch Cave (V10C; Fig. 5.7). Data from Culver (1987).

Differences among Spring Populations

Relative to cave populations, spring populations showed less amongsite variation. There was no significant difference in allometric coefficients for the relationship of eye area and head length. Taken together, the resurgence populations had an allometric equation of

$$\text{Eye Area} = 0.65(\text{Head Length})^{1.55} \qquad (5.6)$$

Table 5.4. The proportion of specimens of *G. minus* retaining various elements of the visual system. All specimens in all spring-dwelling populations examined possess all the optic elements listed. Population codes are given in Figs. 5.7 and 5.8. N = sample size. From Dai (1989).

Optic element	Population			
	VII27C (N = 4)	VI13C (N = 7)	III4C (N = 10)	II21C (N = 6)
Two chambers	1.00	0.00	1.00	1.00
Crystalline cones	0.25	0.00	0.60	0.67
Retinular cell projections	1.00	0.00	0.50	1.00
Reflecting pigment cells	1.00	0.00	0.20	0.50
Retinular and glial cells	1.00	1.00	1.00	1.00
Optic ganglion	0.00	0.00	0.00	0.00

There were significant differences in allometric coefficients of pigment-bundle number with head length, primarily because of a (positive) coefficient of 1.50 for the Organ Cave resurgence population. All other resurgence populations showed a negative allometric relationship between head length and ommatidia number, with b ranging from 0.45 to 0.85 (Culver 1987). The greater similarity of eyes among spring populations and the greater similarity of the equations describing morphometric relationships among spring populations could result from migration among springs, which we argued earlier in this chapter is likely. The similarities could also result from strongly convergent selection pressures, selection that is more convergent in springs than in caves. In Chapter 6 we argue that this is unlikely.

Given the overall difficulty in distinguishing one spring population from the other morphologically, at least on the basis of eyes, it is not surprising that there is no detectable trace of correlation between the eye structure of cave populations and that of populations in their resurgences. For example, the cave population with the smallest allometric coefficient among cave populations does not resurge at the spring whose population has the smallest allometric coefficient among spring populations.

Overall Morphological Patterns

In all resurgence populations but not in all cave populations studied by Culver (1987), body size had a significant effect on ommatidia number. However, any biases introduced are likely to be in the direction of minimizing cave and spring differences since spring populations tend to have smaller body size. There is nearly a twenty-fold change in ommatidia number from the population with the smallest number of ommatidia (Organ Cave, 1812 Stream) and the population with the largest number of ommatidia (Davis Spring). It is apparent that most of the differentiation is by habitat rather than by basin. There were significant differences in ommatidia number (Table 5.5) among cave, karst-window, and resurgence populations (Kruskal-Wallis statistic = 16.766, $p < 0.001$). Median ommatidia number in cave populations was only 1.0 while it was 24.8 in resurgence populations. Median ommatidia number in karst-window populations (21.7) approached that of resurgence populations.

Genetic versus Morphological Patterns

One of the hallmarks of *Gammarus minus* is the strikingly contrasting patterns of genetic and morphological variation. While even a cursory examination of the data indicates that this is the case, the contrast is worth examining in some detail because each pattern is itself correlated with either habitat type (morphological pattern) or hydrologic connectivity (genetic pattern). The evidence presented by *G. minus* concerning these contrasting patterns is what makes the species particularly useful as a model for understanding adaptation.

First we compared genetic and morphological patterns with the hydrology of the karst basins studied. There are six subsurface basins and a set of resurgences (see Fig. 5.7), so seven genetic and morphological clusters were formed for comparison with the seven hydrological clusters. *K*-means clustering (MacQueen 1967) assigns each observation to the clusters with minimum distances to their mean vectors. The seven clusters for morphology are shown in Table 5.6.

Table 5.5. Summary of ommatidia number for mature male *G. minus* from twenty-four cave (C), karst window (KW), and resurgence (R) sites in West Virginia (see Fig. 5.7 for coding).

Site code	Site name	N	Mean	Standard error
I1R	Bone-Norman resurgence	11	24.8	1.03
I2KW	Taylor Spring	20	27.4	1.09
I3C	Bone-Norman Cave	9	20.0	1.32
III4C	Benedicts Cave	20	2.0	0.43
III5R	Davis Spring	20	32.9	0.88
III6KW	Higginbothams Cave	15	17.1	1.15
III7C	Ludington Cave	15	0.9	0.31
III8KW	Milligan Creek	20	23.8	1.03
V9C	Hofsackers Cave	20	4.3	0.75
V10C	Burnside Branch Cave	20	4.4	0.76
V11R	Dickson Spring	20	27.3	1.18
VI12R	Organ Cave resurgence	20	21.7	0.86
VI13C	Organ Cave—Organ Stream	20	0.70	0.26
VI14C	Organ Cave—1812 Stream	20	0.4	0.15
VI15C	Organ Cave—Sively No. 3	20	0.65	0.18
VI16C	Organ Cave—Hedricks	10	0.80	0.42
VI17C	Organ Cave—Masters	10	0.70	0.34
VI18C	Organ Cave—Big Canyon	14	1.0	0.35
II19R	Burns Cave No. 2	20	27.6	0.90
II20R	Spring Creek, Blue Hole	20	23.1	0.71
II21C	The Hole Cave	20	1.0	0.32
IV22KW	Boyd Spring	10	19.6	0.90
IV23R	Scott Hollow resurgence	22	24.2	1.42
IV24C	Scott Hollow Cave	14	6.0	1.00

For electrophoretic data, *k*-means clustering could not be used since only similarity (dissimilarity) matrices were available. In this case, clusters were determined from a UPGMA dendrogram (Table 5.7).

For genetic clusters, there is considerable concordance with hydrologic clusters. The agreement statistic (Cheverud 1982),

$$K = \frac{\sum p_{ii} - \sum p_{i+} p_{+i}}{1 - \sum p_{i+} p_{+i}} \tag{5.7}$$

Table 5.6. Summary of seven population clusters of *G. minus* based on *k*-means clustering of ommatidia numbers (Table 5.5). Refer to Fig. 5.7 for coding of localities.

Cluster description	Cluster composition
1. Caves in basins II, III, and VI	II21C, III4C, III7C, VI13C, VI14C, VI15C, VI16C, VI17C, VI18C
2. Caves in basins IV and V	IV24C, V9C, V10C
3. Cave, karst window, and resurgence	I3C, IV22KW, VI12R
4. Karst window	III6KW
5. Resurgence	III5R
6. Karst window and resurgences	I2KW, II19R, V11R
7. Karst window and resurgences	I1R, II20R, III8KW, IV23R

Table 5.7. Summary of seven population clusters of *G. minus* based on UPGMA dendrogram of electrophoretic distance (Fig. 5.10). Refer to Fig. 5.7 for coding of localities.

Cluster description	Cluster composition
1. Resurgences plus basin I	I1R, I2KW, I3C, II19R, II20R, III5R, IV23R, V11R, VI12R
2. Basin IV	IV22KW, IV24C
3. Basin V	V9C, V10C
4. Basin VI	VI13C, VI14C, VI15C, VI16C, VI17C, VI18C
5. Part of basin III	III8KW
6. Part of basin III	III4C
7. Basin II and part of III	II21C, III6KW, III7C

where p_{ii} is the proportion of cases along the diagonal of the 24 × 24 contingency table, and where + indicates non-i, was high (0.739), with a theoretical range of 0 to 1. Resurgence populations cluster as a distinct group. Also, with the exception of populations in basin I, cave populations are genetically differentiated from resurgence populations. Finally, among cave populations, clustering is by basin, with one exception. The exceptional case is basin III, the hydrology of which was discussed earlier in the chapter.

Seven morphologic clusters determined by k-means clustering present a strikingly different pattern from hydrologic or genetic clusters. Overall the agreement statistic, K, between morphologic and hydrologic clusters was 0.383. Cave populations of basins II, III, and VI form a large cave cluster (Table 5.6). Cave populations from basins IV and V, with less eye reduction, form a second cave cluster. The remaining five morphologic clusters are various combinations of resurgence populations, karst-window populations, and a cave population from basin I.

A visual picture of the concordance of genetic and hydrologic clusters and the discordance of morphological and hydrological clusters is given in Figure 5.15. The number of arrows connecting the cluster patterns and the number of times they cross reflect the degree of discordance.

To complete the analysis we carried out an identical procedure on data from the three habitats (caves, karst windows, and resurgences). The three morphologic clusters (Table 5.8) have a concordance with the habitat clusters of 0.656. One cluster corresponds to all but one of the cave populations. A second cluster comprises a single resurgence (Davis Spring), and the third cluster contains the rest (all karst windows, all other resurgences, and one cave population). The one cave population (Bone-Norman) is not differentiated either morphologically or genetically from other populations in the same basin.

The three genetic clusters do not in general correspond to the three habitat clusters. The concordance is lower than was the case for morphology ($K = 0.440$). One cluster corresponds to basin VI (Organ Cave streams); one corresponds to caves and karst windows in basin

114   The Geography of *Gammarus minus*

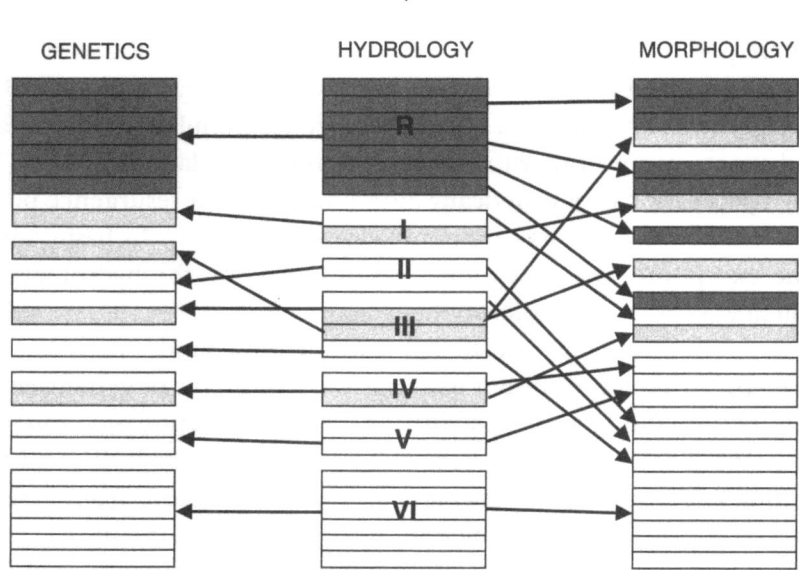

Fig. 5.15. Comparison of seven clusters of *G. minus* populations based on electrophoretic data (labeled GENETICS) and seven clusters based on ommatidia number (labeled MORPHOLOGY) with the six hydrologic basins and resurgences shown in Fig. 5.7. Arrows indicate source of genetic and morphological clusters. Open bars are caves, lightly shaded bars are karst windows, and darker shaded bars are resurgences.

Table 5.8. Summary of three population clusters of *G. minus* based on *k*-means clustering of ommatidia numbers (Table 5.6). Refer to Fig. 5.7 for coding of localities.

Cluster description	Cluster composition
1. Caves in basins II, III, IV, V, and VI	II4C, III7C, V9C, V10C, VI13C, VI14C, VI15C, VI16C, VI17C, VI18C, II21C, IV24C
2. Resurgences, karst windows, and one cave	I1R, I2KW, I3C, III6KW, III8KW, VI12R, V11R, II19R, II20R, IV22KW, IV23R
3. Resurgence	III5R

Table 5.9. Summary of three population clusters of *G. minus* based on UPGMA dendrogram of electrophoretic distance (Fig. 5.10). Refer to Fig. 5.7 for coding of localities.

Cluster description	Cluster composition
1. Basin VI	VI13C, VI14C, VI15C, VI16C, VI17C, VI18C
2. Basin II and part of III	III6KW, III7C, II21C
3. Resurgences plus basins I, IV, V, and part of III	I1R, I2KW, I3C, III4C, III5R, III8KW, V9C, V10C, V11R, V12R, II19R, II20R, IV22KW, IV23R, IV24C

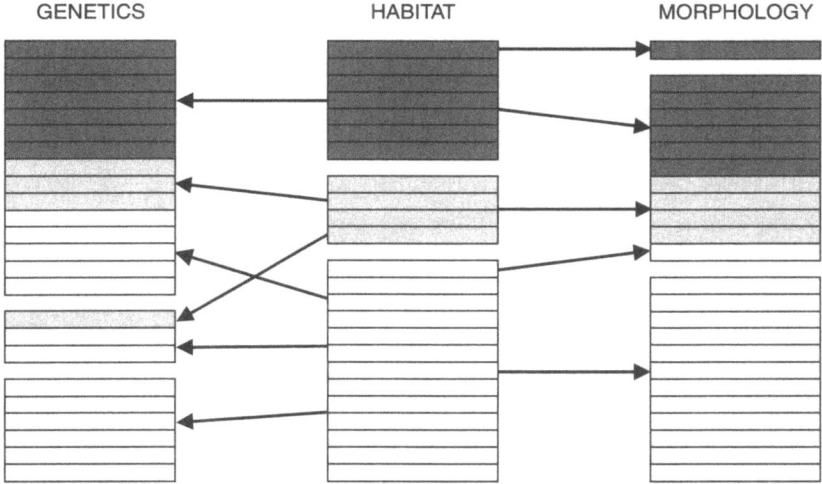

Fig. 5.16. Comparison of three clusters of *G. minus* populations based on electrophoretic data (labeled GENETICS) and three clusters based on ommatidia number (labeled MORPHOLOGY) with the three habitats—caves *(open bars)*, karst windows *(lightly shaded bars)*, and resurgences *(darker shaded bars)*. Arrows indicate source of genetic and morphological clusters.

II and part of basin III; the third comprises the remaining resurgences, karst windows, and caves (Table 5.9). A visual picture of the concordance of morphological and habitat clusters and the discordance of genetic and habitat clusters is given in Figure 5.16.

Summary

Relative to most species in the genus, *Gammarus minus* has a southerly distribution, sharing an affinity for springs and caves with other *Gammarus* with southerly distributions. Although it is found in caves throughout its range, from Pennsylvania to Oklahoma, highly troglomorphic populations are found only in two relatively small regions of Virginia and West Virginia. Troglomorphic populations occur only in extensive cave systems in large subsurface basins with few if any direct surface inputs. Furthermore, the habitable space for *G. minus* is fragmented, as subsurface basins and their resurgences are separated by uninhabitable "phreatic loops." These subsurface basins contain both cave stream habitats and occasional openings to the surface (karst windows), which also have populations of *G. minus*.

Genetic analysis of electrophoretically detectable protein variation indicates that differentiation is largely determined by hydrological relationships among populations. Cave and karst-window populations within a basin tend to cluster together, and resurgence populations tend to cluster together. Clustering is by basin, not by habitat. F-statistics suggest little if any gene flow between basin and resurgence and little if any gene flow among populations of different basins. Most populations have heterozygosity levels typical of aquatic invertebrates, and there is no evidence for genetic bottlenecks.

The extensive morphological variation in *G. minus* shows a very different pattern. Variation in eye size was the most thoroughly analyzed. In this case, populations clustered by habitat rather than by basin. Cave and resurgence populations are strikingly different. Karst-window populations exhibit a wide range of eye sizes, from very small (a highly troglomorphic character) to the sizes seen in resurgence populations. Eye size varies as well among cave populations, but to a lesser extent among resurgence populations.

Patterns of eye-size variation among cave populations and the genetic structure of populations strongly suggest independent invasion of subsurface basins. On the other hand, the overall similarity of eyes in cave populations relative to resurgence populations points to a role for natural selection, which is the focus of the next chapter.

Selected References

Barnard, J. L., and C. M. Barnard. 1983. *Freshwater Amphipoda of the world.* 2 vols. Mount Vernon, Va.: Hayfield Assoc. A comprehensive albeit somewhat idiosyncratic treatment of the historical biogeography of *Gammarus* and other amphipods.

Culver, D. C. 1987. Eye morphometrics of cave and spring populations of *Gammarus minus* (Amphipoda: Gammaridae). *Journal of Crustacean Biology* 7:136–147. A detailed eye-morphometric study to date of *G. minus.*

Ford, D. C., and P. W. Williams. 1989. *Karst geomorphology and hydrology.* London: Unwin Hyman. A thorough discussion of karst, karst basins, and karst hydrology.

Jones, W. K. 1973. *Hydrology of limestone karst in Greenbrier County, West Virginia.* Bulletin 36. Morgantown, W.Va.: West Virginia Geologic and Economic Survey. A detailed account of the hydrology of the main study area.

Sarbu, S., T. C. Kane, and D. C. Culver. 1993. Genetic structure and morphological differentiation: *Gammarus minus* (Amphipoda: Gammaridae) in Virginia. *American Midland Naturalist* 129:145–152. A detailed study of geographical genetics of the Ward's Cove *G. minus* populations.

Weir, B. S. 1990. *Genetic data analysis.* Sunderland, Mass.: Sinauer Associates. A thorough account of *F*-statistics.

6 Making a Case for Selection

▼

Establishing that natural selection is responsible for the evolution of an adaptive trait requires, at a minimum, a demonstration that variation in the trait being considered is genetically determined, at least in part, and that differences in the trait result in differences in reproductive success. In this chapter we lay out the evidence for the heritability of the troglomorphic traits noted in *Gammarus minus*—in particular, the elaboration of antennae and the reduction of eyes—and the evidence for directional selection producing these traits.

We must note, however, that two other factors play a potentially major role in morphological evolution. First, it is possible that non-linear selection, such as stabilizing selection, is also at work. We investigate non-linear selection using splines, a type of non-parametric regression that allows an objective evaluation of the shape of the fitness function. Second, any explanation of adaptation may be complicated by the fact that genetic correlations between traits can be both constraints on selection and the object of selection. We use the graphical technique of biplot (Gabriel 1971) to clarify the patterns of genetic correlation among habitats and among basins.

Heritability

Although there is considerable genetic differentiation among *G. minus* populations, especially between resurgence and cave populations and among cave populations in different basins (Chapter 4), we may not automatically conclude that the differences among populations in troglomorphic characters and the differences in troglomorphic characters among individuals within a population are genetically determined. It is especially important to consider other causes with respect to changes in cave animals, since many apparently troglomorphic features could be environmentally determined. Starving animals in general have a "more delicate appearance"—the description used by Shoemaker (1940) for the highly modified *tenuipes* populations of *G. minus* as well as by Racovitza (1907) for cave animals of all types.

Fong's heritability studies (1985, 1989) largely confirmed that the features observed by Shoemaker had a genetic basis. In a laboratory breeding experiment using animals from two cave populations (III4C and VI13C in Fig. 5.7) and two resurgence populations (III5R and VI12R in Fig. 5.7), Fong found that, for the most part, the morphological differences observed between cave and spring populations are maintained in the offspring raised under identical conditions of constant darkness at 15°C. In both cave populations, all six antennal characters measured were larger than in the resurgence populations from the same basin (Table 6.1). The percent increases ranged from 8 percent (second antennal peduncle length in basin III) to 47 percent (first antennal flagellum length in basin III).

Likewise, offspring of cave animals reared under identical conditions with the offspring of spring animals had smaller eyes. In this case the difference was even more striking. Reductions ranged from 68 percent for eye area in basin III to 84 percent for ommatidia number in basin VI (Table 6.1). Only in the case of head length, a measure of overall size, was there a discrepancy in the direction of change between cave and spring populations. In basin VI, the spring population was larger under constant laboratory conditions (Table 6.1) while the cave population was larger in the field (Holsinger and Culver 1970). This discrepancy has its roots in the nature of selection on body size in resurgence populations, a topic we take up later in the chapter.

As Lewontin (1974) points out, even if all the among-population differentiation is genetic, this implies nothing about within-population

Table 6.1. Trait means and coefficients of variation (C.V.) for sixty-day-old *G. minus* from a resurgence and cave population from each of two basins (see Fig. 5.7): Davis Spring (III5R); Benedicts Cave (III4C); Organ Spring (VI12R); and Organ Cave (VI13C). Traits are head length (hl), eye ommatidia number (en), surface area of compound eye (ea), peduncle length of first antenna (p1), flagellum length of first antenna (l1); number of flagellum segments of first antenna (n1), and the second-antenna analogs (p2, l2, and n2). N = sample size. From Fong (1989).

	Population							
	III5R (N = 182)		III4C (N = 137)		VI12R (N = 299)		VI13C (N = 80)	
Trait	Mean	C.V.	Mean	C.V.	Mean	C.V.	Mean	C.V.
hl[a]	4.48	12.6	4.65	11.1	4.96	11.6	4.46	10.1
en	18.31	22.6	5.14	26.2	19.67	18.5	3.22	31.7
ea[b]	14.85	26.2	4.70	28.8	16.61	23.8	4.32	24.7
p1[a]	5.40	19.6	6.28	14.5	5.06	14.0	6.18	16.6
l1[a]	12.65	22.7	18.64	16.0	13.05	15.1	15.35	20.3
n1	11.10	17.4	14.00	13.4	10.34	13.5	12.87	16.0
p2[a]	4.91	19.1	5.30	14.9	4.22	14.7	5.39	18.8
l2[a]	4.02	21.0	4.90	16.5	3.54	17.4	4.59	21.2
n2	4.57	18.5	5.10	15.3	4.12	8.7	5.15	19.8

a. In mm × 10.
b. In mm² × 1,000.

variation. It is ultimately the within-population variation that concerns us, since our goal is to understand the process of differentiation and adaptation, not just the end result. Coefficients of variation, which reflect within-population variation, for traits in the four populations in Fong's studies ranged from 8.7 percent to 31.7 percent. Coefficients of variation were generally highest for the eye traits in all four populations. The question is whether this variation has a genetic component.

The measurement of heritability, the percentage of overall phenotypic variation in a trait due to additive genetic variation, has a long history of use in applied genetics, particularly in connection with artificial selection (Falconer 1989). Evolutionary biologists have increasingly used heritability and the related measurement of genetic correlation in large part because of the central role they play in Lande and Arnold's predictive theory of phenotypic evolution (1983). On the other hand, the use of heritability and genetic correlation in evolutionary theory has serious conceptual problems, most important of which is the artificial separation of organism and environment (Levins and Lewontin 1985). Heritability and genetic correlation depend on the range of genetic variation and on the range of environments in which it is measured. Finally, there are considerable statistical problems associated with the measurement of heritability and genetic correlations (Shaw 1987). The statistical problems are exacerbated for *G. minus* (and, we suspect, for most populations in nature) because we were unable to obtain unbiased measures of heritability. The most accessible method for measuring heritability in *G. minus* is to utilize differences between groups of full-sibs. Full-sib analysis results in a measure of broad-sense heritability (Falconer 1989):

$$\frac{V_A + V_D/2 + 2V_{Ec}}{V_A + V_D + V_{Ec} + V_{Ew}} = h^{2*} \qquad (6.1)$$

where the V_A is the additive genetic variance, V_D the variance due to dominance deviations, V_{Ec} variance due to common environment of sibs (for example, maternal effects), and V_{Ew} other environmental effects. The denominator of equation (6.1) is the total phenotypic variance. Preferred experimental designs, such as half-sib analysis or parent-offspring regression, yield the desired narrow-sense heritability estimate:

$$\frac{V_A}{V_A + V_D + V_{Ec} + V_{Ew}} = h^2 \qquad (6.2)$$

This unbiased heritability estimate could not be obtained for *G. minus* because, at least in the laboratory, males would not mate repeatedly and as a result few half-sibs were available for measurement.

Given the general problems with interpreting heritabilities and the specific problems with measuring heritability in *G. minus*, why is it worth pursuing? First, while significant heritabilities from equation

(6.1) might indicate only significant maternal effects, non-significant effects indicate heritabilities not different from zero. It is important to establish at least the possibility of heritable variation. Second, we were able to take steps to minimize the likely magnitude of V_{Ec}. In a study of the ontogenetic changes in the estimate of heritability from equation (6.1), Fong (1985) found a reduction until 60 days of age, likely the result of a diminution of maternal effects on heritabilities. Therefore, we measured animals at 60 days of age. Third, full-sib analysis allows not only an estimate of heritability but also an estimate of genetic correlations. Genetic correlations are an important tool in the analysis of evolutionary tradeoffs, an issue at the heart of the evolution of *Gammarus minus*.

Heritability estimates for head length, a measure of overall size, six antennal characters, and two eye characters are given in Table 6.2. Because of the correlation of body size and most eye and antennal characters and the increase in body size with age, care was taken to eliminate size (and age) effects from the analysis. To this end, residuals of the regression of the antennal characters and eye characters on head length were used in the analysis. Overall, broad-sense heritabilities were large, with a mean of 0.68, and all but two of thirty-six were statistically significant.

Even taking into account a potentially large maternal effect, heritabilities are large. Heritabilities for body size (measured as head length) averaged 1.09, indicating that maternal effects must be present. But even if V_{Ec} were equal to V_A, a large value even for mammals (see Falconer 1989, p. 172), and other variance components relatively small, then V_A would be quite large, approximately 0.7. Heritabilities for eye characters averaged 0.7 and heritabilities for antennal characters averaged 0.6. If maternal effects are largely reflected in overall size, perhaps not unlikely considering the potential size constraints of the brood pouch, eye and antennal heritabilities are independent of this effect since they were calculated using the residuals of regression on body size. Except for body size, heritabilities were generally higher (fourteen of sixteen cases) for cave populations than for resurgence populations in the same basin (Table 6.2).

Table 6.2. Trait broad-sense heritabilities (h^2) and standard errors (S.E.) for sixty-day-old *G. minus* from a resurgence and cave population from each of two basins (see Fig. 5.7): Davis Spring (III5R); Benedicts Cave (III4C); Organ Spring (VI12R); and Organ Cave (VI13C). Traits are head length (hl), eye ommatidia number (en), surface area of compound eye (ea), peduncle length of first antenna (p1), flagellum length of first antenna (l1); number of flagellum segments of first antenna (n1), and the second-antenna analogs (p2, l2, and n2). From Fong (1989).

	Population							
	III5R		III4C		VI12R		VI13C	
Trait	h^2	S.E.	h^2	S.E.	h^2	S.E.	h^2	S.E.
hl	1.12	0.16	1.07	0.19	0.85	0.12	1.34	0.26
en	1.08	0.16	0.97	0.19	0.46	0.11	0.79	0.27
ea	0.42	0.15	1.22	0.19	0.11[a]	0.10	0.79	0.27
p1	0.69	0.16	1.37	0.19	0.50	0.11	0.57	0.26
l1	0.51	0.15	1.06	0.19	0.50	0.11	0.91	0.27
n1	0.50	0.15	0.84	0.19	0.37	0.11	0.31[a]	0.26
p2	0.46	0.15	0.71	0.18	0.63	0.12	1.18	0.26
l2	0.59	0.16	0.61	0.18	0.68	0.12	0.86	0.27
n2	0.30	0.14	0.48	0.18	0.38	0.11	0.86	0.27

a. Not significant.

Evidence for Directional Selection

Since cave and resurgence populations are clearly differentiated with respect to overall size, eyes, and antennae, and since heritable variation in these characters is indicated, the obvious next step in our examination of adaptation is to look for evidence that natural selection, acting on the within-population variation, could have produced the observed changes.

An estimation of the total selection on characters requires a measure of lifetime fitness, including fecundity and mortality. Although a lifetime fitness measure was not possible with *G. minus*, two components of lifetime fitness are easily analyzed:

1. Amplexus—the carrying of females by males prior to fertilization (see Chapter 5)
2. Fecundity—the number of fertilized eggs carried in the external brood pouch by ovigerous females

Both of these components of fitness may be affected by changes in troglomorphic characters. The ability to find mates as well as food may be linked to the quality of extra-optic sensory structures, such as antennal length. In turn, the elaboration of extra-optic sensory structures at the neurological level may require a diminution of optic structures at the neurological level. The inverse relationship in relative size of optic and olfactory lobes of *G. minus* (Chapter 3) is consistent with this hypothesis.

If, in a cross-sectional analysis of selection as is summarized here, the sample contains individuals of different ages, then the change of a character with age could be misinterpreted as selection. For example, the observation that increased body size is correlated with mating success could be the result of older individuals tending to mate more, or it could be the result of selection for increased size independent of age. We excluded cases that could be the result of age alone rather than selection for size by considering only sexually mature individuals and by eliminating the effect of size from all variables by utilizing the residuals of the regression against head length. Because the expected patterns of selection on different groups of characters (body size, antennae, and eyes) are different, synthetic characters such as those generated by principal-components analysis were not used.

With the use of multiple regression models, selection can be measured on a suite of characters in a way that takes into account any phenotypic correlations between them (Lande and Arnold 1983). In our study of directional selection (Jones, Culver, and Kane 1992), we employed the following characters:

1. Head length, a measure of body size
2. Eye area and ommatidia number, measures of optic structure
3. Peduncle lengths of first and second antennae and number of flagellar segments, measures of extra-optic sensory structure

The standardized directional selection gradients, B, for the multiple regression

$$W_R = a + \sum B_i X_i \qquad (6.3)$$

measures the correlation between fitness, W_R (either fecundity or amplexus), and a character, X_i (standardized head length and residuals of eye and antennal characters on head length), when all other characters are held constant. If the characters included in the model are phenotypically uncorrelated with any unmeasured character, then B is a measure of the direct effect of selection on this character. Computational details are in Jones, Culver, and Kane (1992).

Directional selection gradients were computed for forty collections of *Gammarus minus* from two caves and three resurgences. Overall, 49 percent of all multiple linear regressions (equation 6.3) and 16 percent of measured selection gradients were statistically significant. Selection gradients are summarized in Table 6.3.

In both caves and resurgences, selection gradients were positive for both head length and antennal characters. Antennal selection gradients tended to be small, averaging 0.06 in both caves and springs. In contrast, directional selection on eyes tended to be negative in caves

Table 6.3. Means, standard deviations (S.D.), and skew of selection gradients (B) for antennal, eye, and head-length characters for *G. minus* from cave and spring habitats. Asterisks indicate means and skews significantly different from zero. N = sample size. From Jones, Culver, and Kane (1992).

Character	Habitat	N	Mean	S.D.	Skew
Head length	Cave	55	0.18**	0.19	0.37
Head length	Spring	36	0.19**	0.22	0.52
Antennae	Cave	130	0.06*	0.21	0.72**
Antennae	Spring	90	0.06*	0.23	0.37
Eye	Cave	85	−0.08**	0.21	−1.09**
Eye	Spring	54	0.06**	0.22	0.54

*$p < 0.05$.
**$p < 0.01$.

and positive in springs. All distributions were skewed away from zero (Table 6.3), indicating that reversed-sign selection gradients (for example, negative selection when the mean is positive) were uncommon. A similar pattern obtains if only statistically significant selection gradients are considered (Fig. 6.1 and Table 6.4). All cases of significant selection on head length in cave populations were positive, and there was no difference in selection on head length between cave and spring populations. Most, but not all, cases of significant selection on antennal length were positive, and the range of values was greater (Fig. 6.1). Significant cave eye selection gradients were largely negative and all spring eye selection gradients were positive. A nested ANOVA (Table 6.4) of significant eye selection with caves and springs as groups and particular caves and springs as subgroups (sites) indicated that 94 percent of the variance was due to habitat differences, and 6 percent was due to differences within sites. The habitat effect was highly significant.

We were unable to resolve completely the circumstances under which strong selection was likely. Significant selection was more likely in winter (Jones, Culver, and Kane 1992), but given the at best weak seasonality of the life cycle in caves (Chapter 4), it is difficult to know what to make of this observation. Selection on body size and eye size was more frequent in caves, while selection on body size and antenna size was more frequent in springs.

The pattern of directional selection observed in caves was consistent between the two caves studied (Organ Cave and Benedicts Cave) and consistent with the differences between cave and resurgence populations. There was selection for larger body size, relatively larger antennae, and smaller eyes. Indeed, cave *G. minus* have larger body size, longer antennae, and smaller eyes than spring individuals (Chapter 3 and Table 6.1). The explanation for selection for larger body size and for a relative increase in antennal size is straightforward—this is selection for increased size of the non-visual sensory system. Because of the strong phenotypic and genotypic correlations between head length and antennal characters (Fong 1989), selection for increased head length will result in increased antennal size.

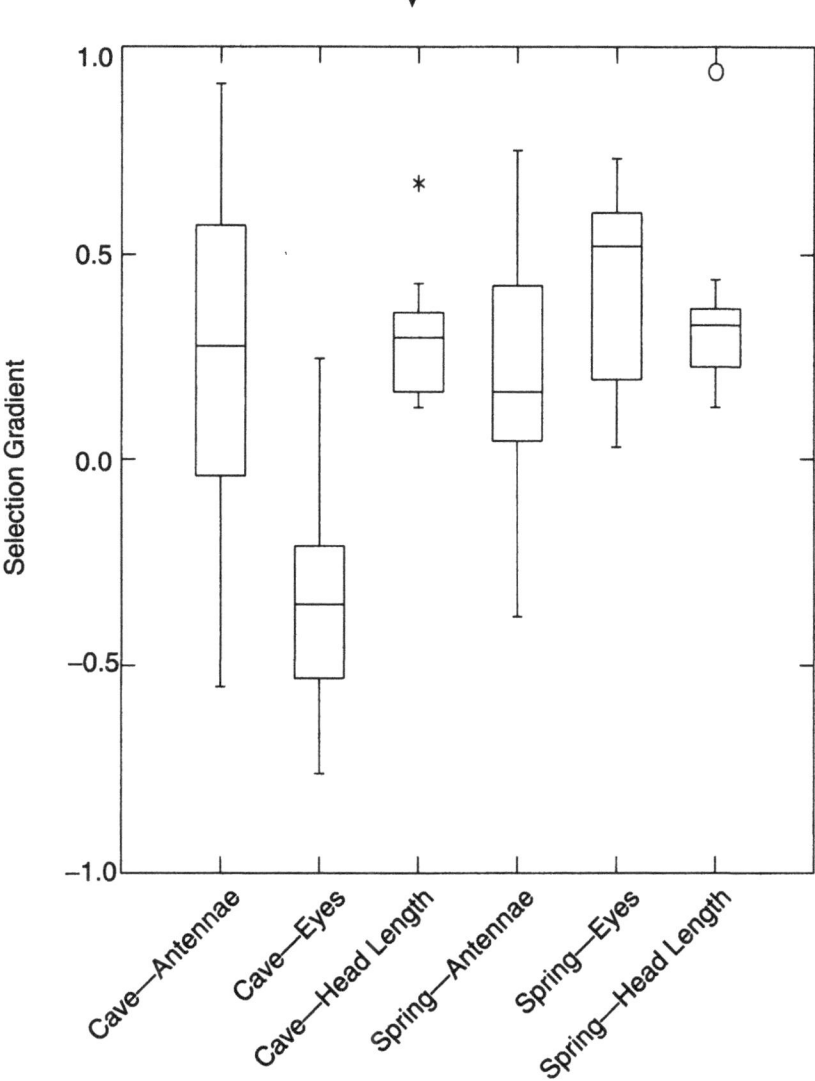

Fig. 6.1. Box-and-whiskers plots of significant selection gradients on standardized log of head length and selection gradients on standardized residuals of log of antennal and eye characters regressed on head length for *G. minus*. The rectangles enclose the interquartile range, the horizontal line is the median, and the "whiskers" mark off a distance of 1.5 times the interquartile range to either side of the box. Asterisks are values falling between 1.5 and 3 times the interquartile range, and circles are values falling more than 3 times the interquartile range, beyond the box. From Jones, Culver, and Kane (1992).

Table 6.4. Nested ANOVA of statistically significant selection gradients on eye characters. Habitats are caves and springs, and sites are particular caves or springs, for example, Davis Spring. F for habitat is $M.S._{Habitat}/M.S._{Sites\ within\ habitat}$. From Jones, Culver, and Kane (1992).

Source	S.S.	d.f.	M.S.	F	p
Habitat	1.487	1	1.487	20.34	.02
Sites within habitat	0.244	4	0.061	0.84	.55
Error	1.019	14	0.073		

Larger antennae result from direct selection on antennae and indirectly from direct selection on head length. An animal with larger antennae has at a minimum a larger radius of perception of its environment. Nonetheless, selection for an increase in relative size of antennae is less frequent and less intense than selection for a relative decrease in eye size (Jones et al. 1992). This difference may be echoed in the observation in several groups of cave animals—for example, cave populations of the Mexican characin fish *Astyanax fasciatus* (Wilkens 1988) and cave populations of various species of hydrobiid snails in the genus *Fontigens* (Hershler, Holsinger, and Hubricht 1990)—that regressed characters are more frequent than elaborated characters.

How can the strong directional selection against eyes be explained? If the advantage of smaller eyes is simply that it makes larger antennae possible, through some form of energy economy or evolutionary tradeoff (Poulson 1963), then this effect should be reflected in direct selection on extra-optic sensory structures such as the antennae. In this case, direct selection on eyes should be negligible. One can always appeal to unmeasured but phenotypically correlated characters (for example, urosome spines), but the characters we have chosen are those for which tradeoffs are most likely. A more plausible explanation is that selection is not acting on any particular part of the anatomy (for example, the first antennal peduncle) but rather on the function it performs (Arnold 1983, Jones and Culver 1989). For antennae and eyes the functions are detection of chemical and visual stimuli, respectively. These two sets of information are not processed independently in the central nervous system of the amphipod, either anatomically or physiologically (Bullock and Horridge 1965). The relative size of *G. minus* brain regions has changed in relation to the relative size of the sensory structures and the amount of stimuli they receive (see Chapter 3). Thus, sensory compensation, a favorite idea of neo-Lamarckians such as Packard, occurs first and foremost at the neurological level in the central nervous system. Apparently, neurological compensation also occurs in cave vertebrates. Voneida and Fish (1984) have shown that in *Astyanax* fish living in surface waters areas of the optic tectum are innervated by the visual sensory system, whereas in *Astyanax* fish living in cave waters the same areas of the central nervous system are innervated by the somatic sensory system. Negative directional selection on eyes and positive directional selection on antennae are both a reflection of selection for neurological function.

The pattern of directional selection observed in springs was similarly consistent among the three springs studied (Davis Spring, Dickson Spring, and Organ Spring), but was not consistent with differences observed between cave and spring populations. Larger eyes were selected for (Table 6.3, Fig. 6.1), and eyes of individuals in spring populations are larger than eyes of individuals in cave populations. However, selection for head length and for relative antennal size was in the same direction as that for cave populations. This was particularly striking in the case of selection for head length. There was significant selection for head length in 39 percent of the cases analyzed, and all were for larger head length. Selection for antennal size was less strong (Table 6.3) and less likely to be statistically significant. Significant selection occurred in only 16 percent of the cases analyzed, and 80 percent of these were for larger relative antennal size.

The most likely explanation for the inconsistency, at least for directional selection for increased body size, lies in the fact that the analysis of selection was a cross-sectional rather than a longitudinal study, and only two components of fitness were considered—mating success and fecundity. *Gammarus minus* does not exist in an ecological vacuum (see Chapter 4), and we must consider the biotic interactions that are likely to have a major impact on mortality patterns. In most of the resurgences where *G. minus* is found its major predator is the sculpin *Cottus carolinensis*. Sculpins, as size-selective predators of amphipods, preferentially eat the larger individuals (Newman and Waters 1984). In a study of size variation among spring populations, Man (1991) found that individuals from sculpin-free springs were consistently larger than individuals from springs with sculpins. For example, ovigerous females from sculpin-free springs were 10 to 30 percent larger than ovigerous females from springs with sculpins (Fig. 6.2). This suggests that cave populations have not evolved toward larger size but that some spring populations, because of differential predation, are subject to mortality selection for smaller size, which counteracts fecundity and mating selection for larger size. A further complication of Man's data is that sculpin-free springs were karst windows while springs with sculpins were resurgences. However, the particular karst windows and resurgences were indistinguishable with regard to temperature and pH (Man 1991), two of the most important environmental parameters (see Chapter 4).

Selection for body size in *Gammarus minus,* as for most organisms, is very complex. The effect of selection for increased size of antennae may be a consequence of selection for overall size. In some springs this selection may be countered by selection for smaller body size due to predation. In common with other *Gammarus* species (see Ward 1988, Elwood and Dick 1990), mate choice in precopula almost certainly has a significant effect on size. In every case that has been examined, size of the male and size of the female in precopula are strongly correlated (Naylor and Adams 1987). In addition, interactions with other species may have an effect. *G. minus* often strongly interacts with other amphipods and isopods in caves (Culver, Fong, and Jernigan 1991). It is often a predator of the isopod *Caecidotea holsingeri,* and the relative size of predator and prey has a major impact on predator success. We can certainly conclude that, unlike antennal size and eye size, overall size is not a troglomorphic character.

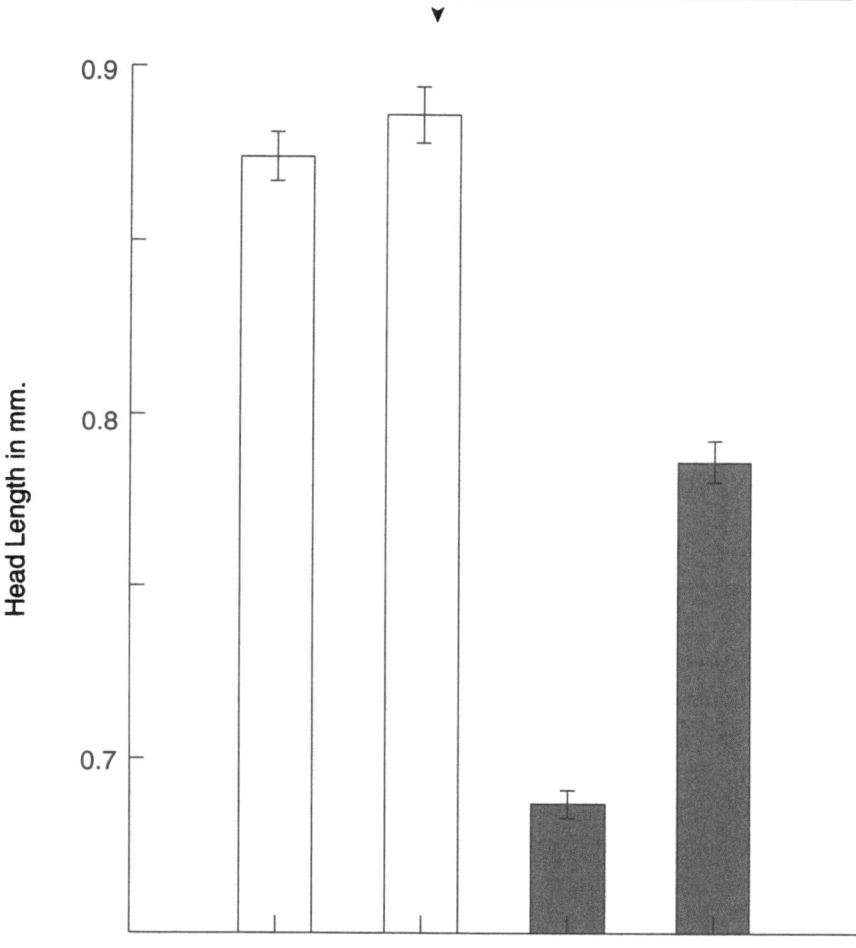

Fig. 6.2. Size, measured by head length, of ovigerous *G. minus* females in springs with sculpin predators *(shaded bars)* and without sculpin predators *(open bars)*. Vertical lines are standard errors. Data from Man (1991).

Evidence for Non-linear Selection

Selection need not be linear or directional (selection for a higher or lower value of a character than its current mean value). Indeed, most evolutionary biologists probably feel stabilizing selection (deviations of a character in either direction from an optimum) has a greater role. In spite of the disparity of views on the importance of directional selection, views on the role of stabilizing selection in cave populations are more consistent. Both Wilkens (1988), arguing from a neutralist view, and Barr (1968), arguing from a neo-Darwinian view, contend that, at least in the early stages of colonization and adaptation to the cave environment, stabilizing selection is reduced. Since eye loss is not complete in cave populations of *Gammarus minus*, these populations are in what is assumed to be the early stages of adaptation. In later stages of evolution in caves, Barr suggests a reconstruction of the "epigenotype"—implying the return of stabilizing selection in cave-adapted populations. Disruptive selection (selection for both extremes of a phenotype) has been little discussed in the context of cave populations. However, models of differential movement and survival in a population continuously distributed from outside to inside a cave (Ludwig 1942, Fong and Culver 1985) at the early stages of colonization imply disruptive selection.

Of the forty collections studied with respect to directional selection (see above), two were of sufficiently large size to thoroughly study possible non-linear components of selection—504 individuals collected from Davis Spring in May 1989 and 658 individuals collected from Organ Cave in December 1988 (Culver et al. 1994). As was the case in the study of directional selection, two fitness components were measured—mating success and fecundity. Five morphological characters were measured—head length; first and second antennal peduncle length; eye area; and ommatidia number. Unlike the directional selection study, characters were not regressed on head length before analysis. Rather, the first three orthogonal vectors obtained from principal-components analysis (PCA) were used. PCA has the advantage of producing orthogonal vectors, thus eliminating the problem of collinearity (Mitchell-Olds and Shaw 1987), and the first vector is almost always a generalized size vector. Indeed, morphometricians sometimes prefer this first PCA to a single morphological measurement as a measure of size (Bookstein 1989). The disadvantage of the artificiality of the variables was partly negated because our focus was on shape rather than on distinguishing selection on different characters.

We did not focus on estimating parameters for linear and quadratic components of selection that can be used to predict long-term phenotypic changes (Lande and Arnold 1983). Rather we focused on obtaining the best estimate of the shape of $f(z)$, which relates individual fitness (w) to the morphological character z under selection:

$$w = f(z) + \varepsilon \qquad (6.4)$$

where ε is a random error term (see Schluter 1988), without making any initial assumptions about what the shape might be. Cubic smoothing splines allow for an objective compromise between a good fit to the data and too much rapid local variation (roughness). Minimizing the modified sum of squares

$$\sum \{z - f(z)\}^2 + n\lambda \int f''(z)^2 \, dz \qquad (6.5)$$

where the first term is the fit to the data and the second term is the roughness penalty, results in a cubic smoothing spline (Silverman 1985). We chose a λ by cross validation, which results in a function f that is the best predictor of points deleted one at a time. Since this procedure makes no *a priori* assumptions concerning the shape of f, it provides an objective estimate of the fitness function (see Schluter 1988). We investigated both the variability of the data around the spline (equation 6.5) that minimizes generalized cross-validation score (GCVS), and the variability of the shape parameter itself (Culver et al. 1994). If $\hat{f}$ denotes the estimates of f from equation (6.2) and f_o the optimal estimate, then (Nychka 1991):

$$\hat{f}(z) - f(z) = [\hat{f}(z) - f_o(z)] + [f_o(z) - f(z)] \qquad (6.6)$$

The second term (variability of data around the spline) was estimated by taking 500 bootstrap samples of the residuals around the curve, using the procedures of Schluter (1988) for fixed λ. Confidence intervals on the shape parameter, λ (the first term in equation 6.6), were determined by taking 100 bootstrap samples of the original data, following Nychka (1991).

Shapes of the fitness function determined by spline fitting allowed us to answer whether most fitness functions in cave and spring populations were linear, monotonic, or with one or more internal maxima or minima. Smoothing splines were calculated for the relationship between amplexus (and fecundity for females) and each of the three principal components using procedures and software provided by Schluter (1988).

Three shape classifications are outlined in Table 6.5—one based on quadratic least-squares regression with confidence intervals obtained by a weighted delete-one jackknife procedure (Mitchell-Olds 1989); one based on optimal splines (equation 6.2); and one based on minimum curvature spline (obtained from the first term in equation 6.3). Of the three cases of apparent stabilizing and disruptive selection, optimal splines indicated that only one was not an artifact of the constraints of least-squares regression. Optimal splines did not in general have multiple internal minima and maxima. Only four of eighteen cases had more than one internal minimum or maximum. Twelve cases were monotonic (either directional selection or no detectable selection). In all but two cases (Table 6.5, Fig. 6.3), confidence intervals of shape included a monotonic function.

In all three classifications, directional selection was more frequent in Organ Cave than in Davis Spring, and non-linear selection was more frequent in Davis Spring than in Organ Cave (Table 6.5). In Organ

Table 6.5. Summary of shape characteristics of selection curves for *G. minus* from Davis Spring and Organ Cave. Numbers are the number of cases out of a maximum of 9 for each population. Upper confidence limit (UCL) spline provides a smoother curve than the optimal spline. Complex shapes have more than one internal maxima and minima. See text for details. From Culver et al. (1994).

Analysis	Shape	Davis Spring	Organ Cave
LEAST-SQUARES	No selection	6	4
	Linear	1	4
	Stabilizing	1	0
	Disruptive	1	1
SPLINE	No selection	3	3
	Monotonic	1	5
	Stabilizing	1	0
	Disruptive	1	0
	Complex	3	1
UCL SPLINE	No selection	6	4
	Monotonic	1	5
	Stabilizing	0	0
	Disruptive	2	0
	Complex	0	0

Cave, there was only one apparent case of non-linear selection, one with several intermediate maxima and minima. This result supports the views of both Barr and Wilkens that stabilizing selection is rare in cave populations, at least in the early stages of adaptation to caves.

In contrast to Organ Cave, there was only one case of simple directional selection in Davis Spring (Table 6.5). We detected cases of non-linear selection on both size and shape parameters. There was a clear case of disruptive selection for overall size (PCA I) for males, and a possible case of stabilizing selection on a shape parameter (PCA III) for ovigerous females. All three PCAs for females show complex selection with multiple maxima and minima. There was at least as much evidence for disruptive selection as there was for stabilizing selection in Davis Spring.

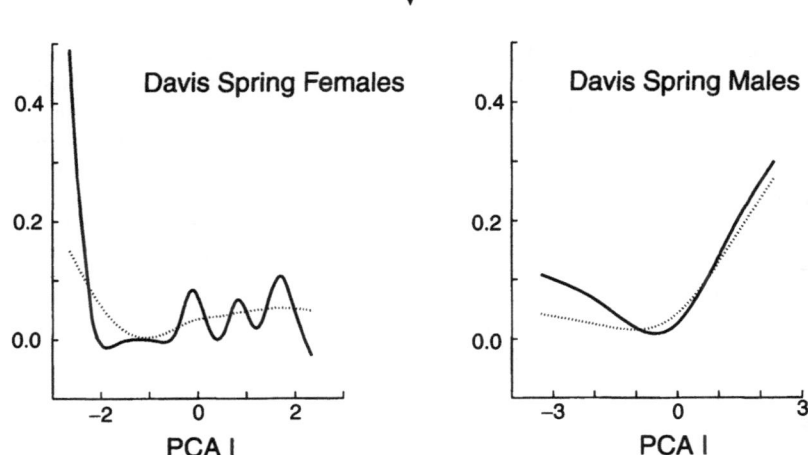

Fig. 6.3. Optimal splines *(solid lines)* and upper confidence limit splines, or minimum curvature splines *(dotted lines)* for fitness for PCA I for Davis Spring females *(left)* and PCA I for Davis Spring males *(right)*. These are the only two cases of eighteen where upper confidence limit splines had an internal minimum or maximum (see Table 6.5). From Culver et al. (1994).

Genetic Correlations

As part of the same experiment described for heritabilities, Fong (1989) estimated genetic correlations among two eye characters (ommatidia number and eye area) and six antennal characters (peduncle length, flagellar length, and number of flagellar segments for each of two pairs of antennae) for each of four populations. These estimates were based on the following relationship:

$$r_g = \frac{cov_b(X,Y)}{[s_b^2(X)\, s_b^2(Y)]^{1/2}} \tag{6.7}$$

where $cov_b(X,Y)$ is the between-family component of covariances between traits X and Y and s_b^2 is the between-family variance component. Table 6.6 shows, in columns, the 28 off-diagonal genetic correlations of each 8×8 genetic correlation matrix from the Organ Cave and Organ Cave resurgence (basin VI) and Benedicts Cave and Davis

Table 6.6. Genetic correlations between eye and antennal characters (rows 1–12), between eye characters (row 13), and between antennal characters (rows 14–28) for *G. minus*. Standard errors are given in parentheses. S = spring population, C = cave population, and numbers refer to the basin. Character abbreviations are en (ommatidia number), ea (eye area), p1 (first peduncle length), l1 (first flagellar length), n1 (first flagellar segment number), p2 (second peduncle length), l2 (second flagellar length), and n2 (second flagellar segment number). Each column represents the listing of the upper triangular of the matrix of genetic correlations. Data are from Fong (1989).

Correlation	S1	C1	S2	C2
1. en–p1	−.33(.15)	.05(.28)	.08(.13)	.68(.06)
2. en–l1	−.21(.16)	.43(.18)	−.21(.14)	.70(.07)
3. en–n1	−.06(.06)	.54(.27)	−.02(.15)	.67(.08)
4. en–p2	−.09(.15)	.32(.18)	.11(.16)	.72(.08)
5. en–l2	−.01(.15)	.42(.18)	−.10(.14)	.41(.04)
6. en–n2	.05(.19)	.41(.19)	−.08(.19)	.68(.10)
7. ea–p1	−.28(.29)	−.09(.28)	.56(.14)	.83(.03)
8. ea–l1	.11(.32)	−.02(.22)	.38(.20)	.68(.06)
9. ea–n1	.34(.32)	.28(.35)	.44(.19)	.52(.10)
10. ea–p2	.20(.28)	.05(.19)	.58(.16)	.56(.10)
11. ea–l2	.42(.23)	−.01(.22)	.32(.19)	.28(.14)
12. ea–n2	.45(.29)	−.01(.23)	.08(.29)	.16(.16)
13. en–ea	.43(.27)	.26(.22)	.72(.09)	.64(.07)
14. p1–l1	.75(.07)	.67(.14)	1.02(—)	.88(.02)
15. p1–n1	.67(.10)	−.02(.44)	.76(.08)	.73(.06)
16. p1–p2	.72(.07)	.95(.02)	.90(.04)	.89(.03)
17. p1–l2	.69(.07)	.83(.08)	.75(.08)	.71(.07)
18. p1–n2	.65(.11)	.25(.25)	.61(.15)	.69(.08)
19. l1–n1	.95(.02)	.77(.14)	.62(.13)	.84(.04)
20. l1–p2	.89(.03)	.85(.05)	1.03(—)	.93(.02)
21. l1–l2	.90(.03)	.87(.05)	.94(.02)	.74(.07)
22. l1–n2	.58(.12)	.83(.06)	.68(.14)	.84(.05)
23. n1–p2	.85(.05)	.67(.07)	.70(.11)	.83(.05)
24. n1–l2	.84(.04)	.68(.19)	.46(.16)	.69(.09)
25. n1–n2	.74(.08)	.76(.15)	.48(.21)	.96(.02)
26. p2–l2	.95(.01)	.99(.01)	.80(.07)	.86(.05)
27. p2–n2	.66(.09)	.45(.15)	.65(.16)	.86(.06)
28. l2–n2	.70(.08)	.64(.12)	.59(.16)	.98(.01)

Spring (basin III). Overall, genetic correlations were positive, averaging 0.53, with 14 percent negative, and 65 percent were significantly ($p < 0.05$) different from zero.

The idea that the genetic variance-covariance matrix and the genetic correlation matrix is of fundamental importance in understanding phenotypic evolution has had wide currency in the past decade, especially as it relates to developmental constraints and evolutionary tradeoffs (see Lande 1982, Cheverud 1982). Perhaps the best-analyzed cases of tradeoffs are life-history characters, such as tradeoffs between survival and reproduction or early and late reproduction. In these cases, one would expect negative pleiotropic effects between the two fitness components. In the case of the evolution of troglomorphy, the selectionist argument is essentially one of negative pleiotropic effects between the two morphological components, coupled with directional selection for increases in the size and the degree of elaboration of extra-optic sensory structures.

Whether the genetic variance-covariance matrix reflects the pleiotropic constraints is another question (Houle 1991). Lande (1982) suggested that negative genetic covariances (and correlations) are likely because alleles that have positive effects on both fitness components will sweep through the population, alleles that have negative effects on both will be rapidly lost, and alleles that have contrasting effects on fitness components will tend to remain at intermediate frequencies longer. However, the prediction of negative covariances has not been borne out for either life-history traits or for troglomorphy. A variety of reasons have been suggested for the failure of tradeoffs to be reflected in negative genetic covariances, including the demonstration by Houle (1991) that positive genetic correlations may be expected even if there is a tradeoff among life-history traits. Houle utilizes the distinction between alleles that affect acquisition of resources and result in positive genetic correlations among traits and alleles that affect allocation of resources and result in negative genetic correlations among traits. If the ratio of acquisition loci to allocation loci is high, positive genetic correlations are more likely. The details of the selectionist model of evolution of troglomorphy differ from that of the evolution of life-history tradeoffs. The evolution of troglomorphy seems to involve directional selection in the allocation of resources to eyes and antennae rather than fitness tradeoffs of life-history traits. Nevertheless, Houle's models show that positive correlations are possible at equilibrium for systems involving negative pleiotropic fitness effects.

In the absence of such a tradeoff between optic and extra-optic sensory structures, eye loss is due solely to the accumulation of structurally reducing, selectively neutral mutations. In this situation there is no reason to expect that the pattern of genetic correlations in the ancestral spring populations themselves resulting from stabilizing selection or mutation-selection balance (Barton 1990) will be changed in the derived cave populations. Similar patterns of correlation, either positive or negative, should occur across habitats within basins but not within habitats across basins.

The final contrast between the selectionist and neutralist hypotheses is the character set in which consistent patterns of genetic correlation are expected to occur. Selection should affect eye-antennal correlations only while neutral mutation could affect all correlations.

We have not focused on whether genetic correlations are positive or negative in a particular habitat such as caves, but whether there are consistent patterns, either positive or negative, in cave populations. If selection molded genetic correlations in either a positive or a negative direction, there should be a consistent pattern. If, on the other hand, selection played no direct or indirect role in the loss of eyes, then no such habitat-specific patterns should occur, and only basin-specific patterns should be present.

For graphical display and interpretation of genetic correlation data, we used a technique called biplot, developed by Gabriel (1971). The biplot is a graphical procedure that displays simultaneously the multivariate structure of both individual observations (genetic correlations) and variables (populations) in a matrix of data. It simultaneously provides a clustering technique (principal-components analysis) for observations, as well as an axis (or scale) for each variable (population) upon which observations may be placed in a manner similar to factor loadings from factor analysis (Jernigan, Culver, and Fong 1994).

The basis of the technique is the factoring of an $n \times p$ data matrix, $\mathbf{Y}$, into a product of two matrices of the form $\mathbf{GH'}$. The factor $\mathbf{G}$ is made up of row vectors $g_1', \ldots, g_n'$, which represent the rows or observations of $\mathbf{Y}$, and the factor $\mathbf{H}$ is made up of column vectors $h_1, \ldots, h_p$, representing of the columns or variables of $\mathbf{Y}$. These vectors are calculated in such a way that the inner product $g_i' h_j$ estimates y_{ij}, the ijth element of the data matrix $\mathbf{Y}$.

The biplot displays the rows (observations), g_i, as a set of points and the columns (variables), h_j, as a set of vectors. In our application the rows correspond to genetic correlations between characters and the columns correspond to the four different populations. Two row points that are plotted close together indicate that they have approximately the same measurements on the entire multivariate set of variables. Two vectors separated by a small (acute) angle indicate that the variables (populations) are highly positively correlated. Similarly, two vectors separated by a large (obtuse) angle indicate that the variables are negatively correlated. For each pair of populations, the element-by-element Pearson correlation coefficient of the genetic correlation matrices was compared against the distribution of this correlation coefficient obtained from all possible permutations of the rows of one population matrix (and corresponding columns)—a procedure known as the Mantel test (see Manly 1991).

All of the within-habitat and within-basin matrix correlations between populations are statistically significant ($p < 0.05$), but none of the correlations that are both between-habitat and between-basin are statistically significant (Table 6.7). That is, populations from similar habitats show significantly similar patterns of genetic correlation among characters, and populations from the same basin also show significantly similar genetic correlation patterns. The maximum correlation was 0.77 between S1 (Organ Spring) and S2 (Davis Spring), and the minimum was 0.37 between S1 and C2 (Benedicts Cave).

The three-dimensional biplot (Fig. 6.4) accounted for 97.2 percent of the variance among rows and 99.8 percent of the variance

Table 6.7. Matrix correlations, probabilities for matrix correlations (based on 5,000 matrix permutations), rank-3 biplot angles (see Fig. 6.5), and cosines of biplot angles (which are rank-3 approximations of the correlations) for each pair of populations. Data are given in Table 6.6. From Jernigan, Culver, and Fong (1994).

	Matrix		Biplot	
Populations	Corr.	Prob.	Angle	Cosine
Same habitat–different basin				
S1–S2	.77	.012	32°	.84
C1–C2	.60	.006	51°	.62
Different habitat–same basin				
S1–C1	.64	.022	55°	.57
S2–C2	.54	.027	46°	.69
Different habitat–different basin				
S1–C2	.37	.097	64°	.44
S2–C1	.44	.072	70°	.35

among columns. The three-dimensional angles are given in Table 6.7. The four three-dimensional vectors form the vertex of a pyramid with a nearly square base. Habitat differences are apparent on the PC I–PC II plane (Fig. 6.5A) and basin differences are apparent on the PC I–PC III plane (Fig. 6.5B). The vertex of the pyramid, which corresponds to [0,0,0], is apparent on the PC II–PC III plane (Fig. 6.5C). A point on the vertex would correspond to the mean genetic correlation for each of the populations, where principal-component scores are zero.

If clusters are formed according to the k-means procedure (MacQueen 1967), up to seven clusters can be formed before the large cluster of thirteen antennal-antennal correlations is broken up. It is the eye-antennal correlations that differentiate the populations because they have different effects on different population vectors.

In the PC I–PC II plane, the two cave vectors (C1 and C2) are very similar and the two spring vectors (S1 and S2) are very similar (Fig. 6.5A). Thus, the effects of habitat on the various genetic correlations can best be understood on this plane. In particular, ea-n2 and en-ea

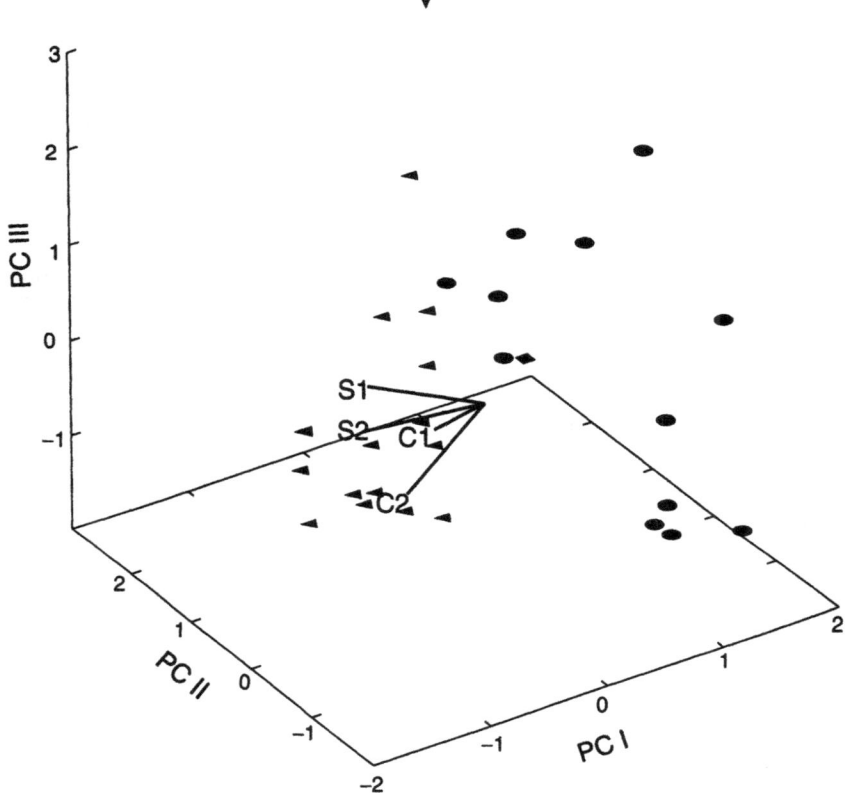

Fig. 6.4. Rank-3 biplot of genetic correlations. Each column of Table 6.6 was standardized by subtracting the column mean and dividing by the column standard deviation; this makes the common vertex in the biplot represent the mean vector. Measurements move through larger values in the direction of the vector. Measurements below the mean are plotted in the opposite direction. Angles are summarized in Table 6.7. Triangles indicate antennal-antennal correlations, circles indicate eye-antennal correlations, and the square indicates an eye-eye correlation. From Jernigan, Culver, and Fong (1994).

correlations (cluster II) have large negative loadings on both cave population vectors and weakly negative loadings on both spring population vectors. The contrasting pattern of strong negative loadings on the two spring population vectors and weakly negative to positive loadings on the two cave population vectors is shown by six eye-antennal correlations (clusters IV, V, and VI). In a sense it is the contrasting projections (loadings) of these eight eye-antennal correlations that keep the angle between spring vectors and the angle between cave vectors small while keeping the angle between vectors representing cave and spring populations in the same basin larger.

In the PC I–PC III plane, the basin 1 vectors (C1 and S1) are very similar and the basin 2 vectors (C2 and S2) are very similar (Fig. 6.5B). Thus, the effects of different basins (different lineages) on the various genetic correlations can best be understood on this plane. The correlations that distinguish the basins are those in the upper center of Figure 6.5B, particularly ea-l1 and p1-n1. Both of these pairs of correlations are larger in basin 1 than in basin 2. Three additional eye-antennal correlations, ea-n1, ea-l2, and en-l1, also have contrasting effects. The contrast with habitat differences is that fewer correlations distinguish basins and that an antennal-antennal correlation is important.

Similarity of genetic correlations within a basin reflects evolutionary history and developmental constraints. If this were the predominant factor promoting similarity among populations, only populations from the same basin should show significant correlations and the biplot of genetic correlations should be similar to the pattern depicted in Figure 6.5B. Since the selective environments of different caves share important features, such as absence of light and reduced food supply, similarity of genetic correlations within a habitat type reflects natural selection. If this were the predominant factor promoting similarity among populations, only populations from the same habitat should show significant correlations and the biplot of genetic correlations should be similar to the pattern depicted in Figure 6.5A. Finally, among the four populations, gene exchange is probably occurring only between the two resurgence populations (Chapter 4). If only ongoing migration were important, only spring populations should be correlated and the biplot of genetic correlations should also be similar to the pattern shown in Figure 6.5B, but with cave population vectors at right angles to each other and to the spring population vectors.

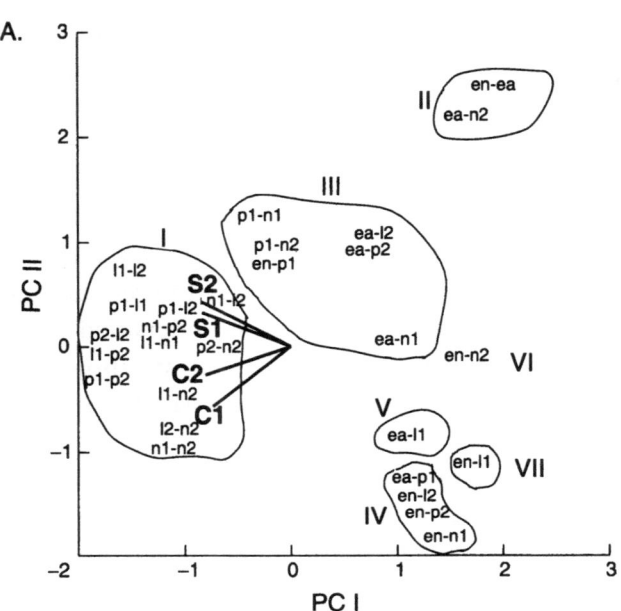
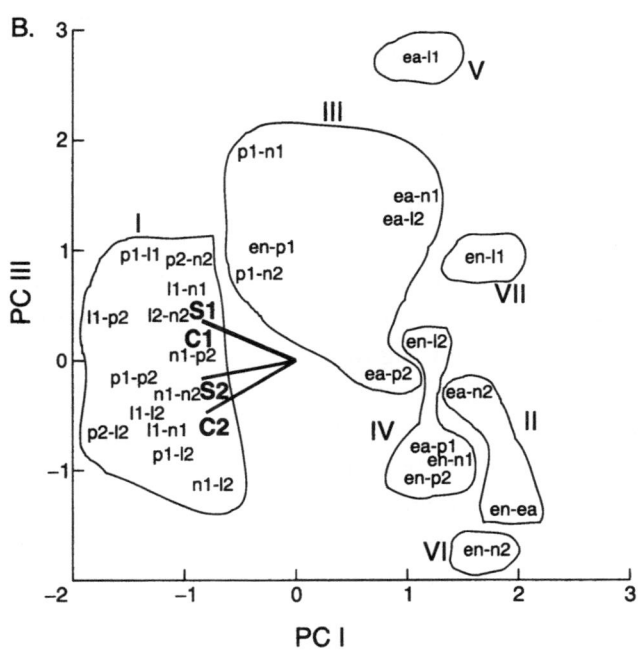

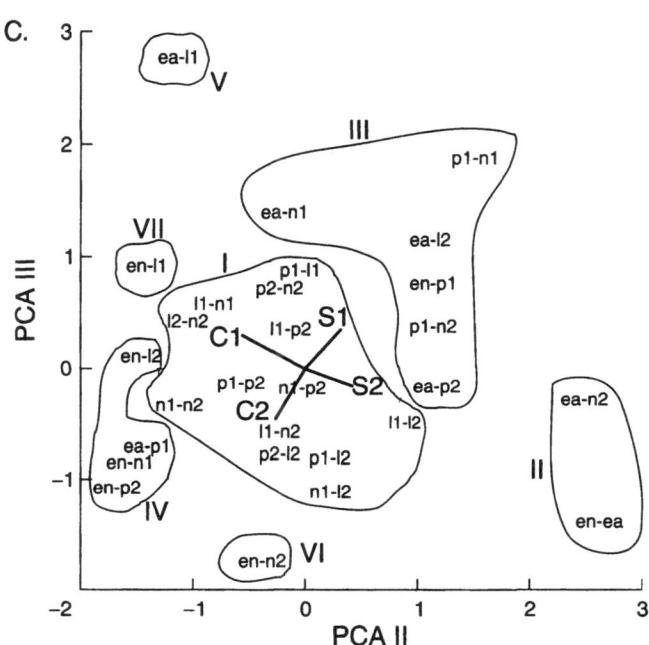

Fig. 6.5. Projections of rank-3 biplot of Fig. 6.4 on *(A)* PC I–PC II plane; *(B)* PC I–PC III plane; and *(C)* PC II–PC III plane. Seven *k*-means clusters are enclosed in circles. Abbreviations for correlations are given in Table 6.6. Cluster I consists of thirteen antennal-antennal correlations. Cluster II consists of an eye-antennal correlation and the eye-eye correlation. Cluster III consists of eye-antennal and antennal-antennnal correlations. Clusters IV, V, VI, and VII are eye-antennal correlations.

In fact, the observed correlations (Table 6.7) and the three-dimensional biplot (Fig. 6.4) are a combination of all three factors. The significant correlation between the two caves indicates the importance of selection; the significant correlations between populations in the same basin indicates the importance of evolutionary history; and the significant correlation between the two spring populations indicates the importance of selection and migration (Table 6.7). Eye-antennal correlations are primarily responsible for the separation of cave population vectors from spring population vectors. These are the correlations that, according to the hypothesis of an evolutionary tradeoff between eye loss and extra-optic sensory structure gain, should consistently differ in cave populations and in spring populations. Both eye-antennal correlations and antennal-antennal correlations are responsible for the separation of basin 1 (basin III in Fig. 5.7) vectors from basin 2 vectors (basin VI in Fig. 5.7). This is in accord with the hypothesis that populations from different basins may differ in any correlation as a result of evolutionary history and not just of eye-antennal correlations, as is predicted by selection models.

The pattern of genetic correlations as displayed by biplot also makes it clear that models of evolution in which the variance-covariance matrix (and genetic correlations) are constant (Lande 1979, Turelli 1988) are insufficient to explain the patterns of genetic correlations, especially eye-antennal correlations. Genetic correlations themselves seem to change in a consistent way during the course of adaptation to the cave environment.

Summary

Differences in eye and antennal sizes between cave and spring populations have a large genetic component. A full-sib analysis of four populations for six antennal, two eye, and one body-size trait indicated significant heritability in nearly all cases, averaging 0.68. Using amplexus and egg number as measures of fitness, significant directional selection on eye, antennal, and body-size characters was observed in 16 percent of the cases. The most striking difference between selection in cave and spring populations was for selection on eyes. In cave populations there was selection for reduced eyes; in spring populations there was selection for larger eyes. This is in accord with the morphological differences between the two populations. The selective advantage of smaller eyes in cave populations is likely to be at the neurological level, where optic connections to the brain may be co-opted by extra-optic sensory systems such as antennae.

In all populations, however, selection tended to favor larger individuals in spite of the fact that individuals from resurgence populations were actually smaller. In resurgences, selection for larger body size in terms of amplexus and egg number may be counteracted by predation by sculpins, resulting in selection for smaller body size in terms of mortality.

Non-linear selection was rare in the cave population studied, and directional selection predominated. The rarity of stabilizing selection in caves accords with the view that there is a release from stabilizing selection during the initial stages of adaptation. In the resurgence population, non-linear selection was more common but it was typically not stabilizing selection. The rarity of stabilizing selection in resurgence populations was surprising, especially because resurgences are presumably the ancestral habitats of the derived cave populations.

Genetic correlations showed significant similarities both between a basin and its resurgence and within a habitat type. Eye-eye and antennal-antennal correlations tended to be similar between a basin and its resurgence, reflecting an ancestor-descendent pattern and thus historical factors. Antennal-eye correlations, those correlations that should reflect evolutionary tradeoffs, tended to be similar within a habitat type, reflecting similar selective regimes in similar habitats. The patterns of genetic correlations thus allow glimpses of the importance of both evolutionary history and selection for adaptive traits.

Selected References

Culver, D. C., R. W. Jernigan, J. O'Connell, and T. C. Kane. 1994. The geometry of natural selection in cave and spring populations of the amphipod *Gammarus minus* Say (Crustacea: Amphipoda). *Biological Journal of the Linnean Society* 52:49–67. A detailed account of estimating non-linear selection in *G. minus* using splines.

Endler, J. A. 1986. *Natural selection in the wild.* Princeton, N.J.: Princeton University Press. A discussion of the various methods of detecting natural selection.

▼

Falconer, D. W. 1989. *Introduction to quantitative genetics*. 3d ed. Essex: Longman. Principles of quantitative genetics, including heritability and genetic correlation.

Fong, D. W. 1989. Morphological evolution of the amphipod *Gammarus minus* in caves: quantitative genetic analysis. *American Midland Naturalist* 121:361–378. Details of heritability and genetic correlation estimates for *G. minus*.

Jernigan, R. W., D. C. Culver, and D. W. Fong. 1994. The dual role of selection and evolutionary history as reflected in genetic correlations. *Evolution,* vol. 48 (forthcoming). Application of biplot to genetic correlation data from *G. minus* populations.

Jones, R., D. C. Culver, and T. C. Kane. 1992. Are parallel morphologies of cave organisms the result of similar selection pressures? *Evolution* 46:353–365. A detailed account of directional selection in *G. minus* populations.

Schluter, D. 1988. Estimating the form of natural selection on a quantitative trait. *Evolution* 42:849–861. The application of splines to estimating selection shapes.

7 Putting the Pieces Together

Presenting the case for *Gammarus minus* as a model for exploring questions of evolutionary biology has required that we marshall evidence from a variety of different fields—the geology and hydrology of karst basins, the ecology of cave and spring habitats, the physiology of our model organism, and statistical studies of morphological and genetic variance in various *G. minus* populations. Now it is time for us to put some of the pieces of this puzzle together. We begin with a summary of the ways that various factors—geography, habitat, hydrology, and genetics—are related to morphology. To accomplish this we use a combination of path analysis and matrix correlation of geographic, habitat, hydrologic, genetic, and morphological dissimilarity. Once we have obtained an overall description of the causal paths among these variables, we proceed to a description of the invasion of and adaptation to caves by *G. minus* in a subsurface drainage basin.

Path Analysis and Matrix Correlation

The influence of common genetic history and natural selection in the evolution of *G. minus* can be cast in terms of causal paths relating physical variables, as measured by geographic, habitat, and hydrologic distances, to the biological variables measured by genetic and morphologic distance (Fig. 7.1). There may be a causal path relating electrophoretic distance with the two possible measures of isolation—geographic and hydrologic distance—due to neutral mutation and genetic drift. For the same reason there may be a separate causal

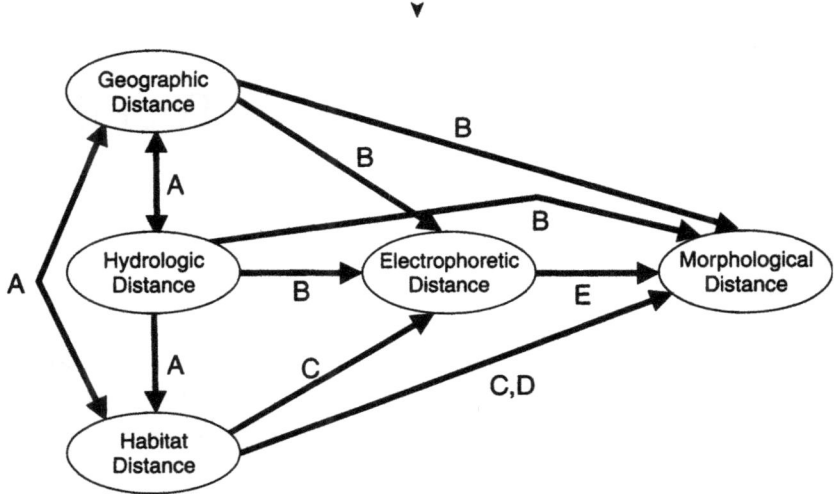

Fig. 7.1. Path diagram showing relationships between various distance measures. A = correlations among non-biological variables; B = causal relationships resulting from isolation; C = causal relationships resulting from selection; D = relationships resulting from environmental effects and gene-environment interactions; and E = causal relationships resulting from different times of colonization. See text for details.

path relating morphologic distance to isolation if the genes measured by electrophoretic techniques are independent and unrelated to genes responsible for morphological differences. On the other hand, habitat differences reflect potential differences in selection, and a path connecting habitat distance (dissimilarity) with electrophoretic distance or morphologic distance indicates natural selection. Finally, the path relating electrophoretic distance with morphologic distance, separately from the direct effects of geographic, hydrologic, and habitat distances, indicates the effects of different times of invasion of different karst basins.

Strong selection in opposite directions on eyes in spring and cave populations (Chapter 6) suggests the habitat distance–morphologic distance path is important. Electrophoretic similarity of populations within basins (Chapter 5) suggests the hydrologic distance–electrophoretic distance path is important. The variation in electrophoretic distance between resurgence and cave populations exhibiting variation in troglomorphic features suggests that the electrophoretic distance—morphologic distance path is important. But are other paths important? For example, geographic distance itself rather than hydrological connection might be important in determining electrophoretic distances. All possible paths are shown in Figure 7.1.

All five variables (geography, hydrology, habitat, electromorphs, and morphology) can be converted to a 24 × 24 matrix of dissimilarities—map distance for geography, Cavalli-Sforza and Edwards arc distance for electromorphs, Euclidean distance for eye morphology, and matrices of 0's and 1's for habitat and hydrology. Habitat distance is set to 0 for comparisons between populations of the same habitat type (caves, karst windows, and resurgences) and 1 otherwise. Hydrologic distance is 0 among cave and karst-window populations within a subterranean basin, 0 among resurgence populations, and 1 otherwise. Each matrix was standardized (mean = 0, S.D. = 1) prior to analysis.

Associations between various distance matrices were used to test correlations and causal relationships among variables, the conceptual equivalent of multiple regression. We used the methodology of Manly (1991), which employs permutation tests of multiple regression of association and allows tests of hypotheses such as those shown in Figure 7.1 (see Douglas and Endler 1982).

Actual matrix correlations among the five variables are shown in Table 7.1. Correlations among the three physical variables (hydrologic, geographic, and habitat distances) are not particularly strong. The largest, between hydrologic and habitat distance, was 0.43. The habitat distance–geographic distance correlation was weak (0.06) and not statistically significant. This pattern makes the task of separating effects of the predictor variables easier. Of the other correlations, the largest are between morphological distance and habitat distance (0.67) and between genetic distance and hydrologic distance (0.64, Table 7.1).

A multiple matrix regression of the effect of the three non-biological distances on electrophoretic distance (Table 7.2) shows a

Table 7.1. Matrix correlations of hydrological distance (BASIN), electrophoretic genetic distance (CAVALLI), geographical distance (GEOG), habitat (HABITAT), and morphological distance (MORPH) on the upper diagonal. On the lower diagonal are significance levels, based on 5,000 matrix permutations. Values in brackets are those determined for cases in which habitat distance between resurgences and karst windows is set to zero.

	BASIN	CAVALLI	GEOG	HABITAT	MORPH
BASIN	—	.639	.204	.426 [.350]	.383
CAVALLI	.000	—	.172	.152 [.194]	.359
GEOG	.001	.011	—	.055 [.004]	.049
HABITAT	.000 [.000]	.032 [.090]	.402 [.951]	—	.670 [.770]
MORPH	.000	.001	.188	.000	—

significant effect of hydrological distance only. The finding that neither geographic distance nor habitat distance influences electrophoretic distance is particularly informative, as it suggests that selection does not determine the pattern of allozyme variation among the populations. The basin effect was very strong, resulting in a partial correlation of 0.69 (Table 7.2).

The second matrix regression of interest is the effect of habitat, genetic, hydrologic, and geographic distance on morphological distance (Table 7.3). Morphological distances are affected by habitat distance and electrophoretic distance, but not by hydrologic or geographic distance, a striking contrast to the pattern for electrophoretic distance. Partial regression coefficients of habitat distance and electrophoretic distance were both significant. The effect of habitat was particularly strong with a partial correlation of 0.67. Both the habitat effect and the genetic effect were significant even if the effect of the other was removed. That is, the residuals of the regression of genetic distances on habitat distances are significantly correlated with morphologic distance ($r = 0.26$, $p < 0.004$), and the residuals of the regression of habitat distances on genetic distances are significantly correlated with morphologic distances ($r = 0.62$, $p < 0.001$). It is just as clear that there is no significant effect of either geographic distance or hydrologic distance on morphologic distance. Partial re-

Table 7.2. Results of matrix regression of hydrologic distance, geographic distance, and habitat distance on Cavalli-Sforza arc distances of electrophoretic distance, following Manly (1991). All data were standardized prior to analysis. Prob1 is the probability of more extreme regression coefficients arising by chance, found using a two-sided test. Prob2 is the percent of randomizations that result in greater additional sums of squares. Prob2 is thus sensitive to order of variables while Prob1 is not. All results are based on 5,000 randomizations. Caves, resurgences, and karst windows are all treated as separate habitats.

Variable	Slope	Prob1	Prob2
Hydrologic distance	0.693	.000	.000
Geographic distance	0.038	.573	.524
Habitat distance	−0.146	.073	.073

Table 7.3. Results of matrix regression of habitat distance, Cavalli-Sforza arc genetic distance, hydrologic distance, and geographic on morphologic distance, following Manly (1991). All data were standardized prior to analysis. Prob1 is the probability of more extreme regression coefficients arising by chance, found using a two-sided test. Prob2 is the percent of randomizations that result in greater additional sums of squares. Prob2 is thus sensitive to order of variables while Prob1 is not. All results are based on 5,000 randomizations. Caves, resurgences, and karst windows are all treated as separate habitats.

Variable	Slope	Prob1	Prob2
Habitat distance	0.666	.000	.000
Genetic distance	0.330	.008	.003
Hydrologic distance	−0.107	.195	.181
Geographic distance	−0.023	.698	.698

gression coefficients are not significant (Table 7.3), and even when considered alone, geographic distance and morphologic distance are not correlated.

We can summarize our findings by replacing the letters on the paths in Figure 7.1 with the corresponding correlations among physical variables and the multivariate regression coefficients on the paths between these variables and genetic and morphologic distances (Fig. 7.2). Despite the fact that hydrologic distance is significantly correlated with both geographic distance and habitat distance (Table 7.1), it alone is significantly correlated with genetic differentiation when all three non-biological distance measures are regressed on electrophoretic distance. Therefore, it is hydrologic distance, a measure of subsurface connectivity, but not geographic distance, a measure of surface connectivity, that is the relevant parameter for understanding the genetic structure of these populations. The lack of correlation between electrophoretic distance and habitat distance suggests that selection does not influence allozyme variation, since it

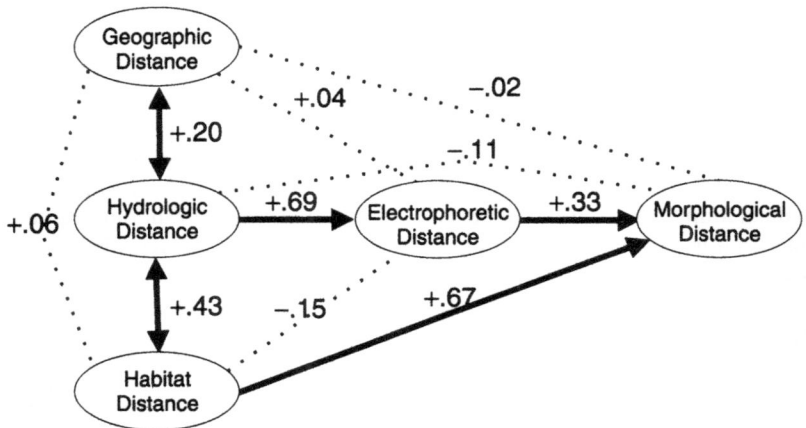

Fig. 7.2. Actual correlations among non-biological variables and multiple regression coefficients between non-biological variables and biological variables (see Fig. 7.1). Dotted lines indicate non-significant effects; solid lines, significant effects.

Putting the Pieces Together

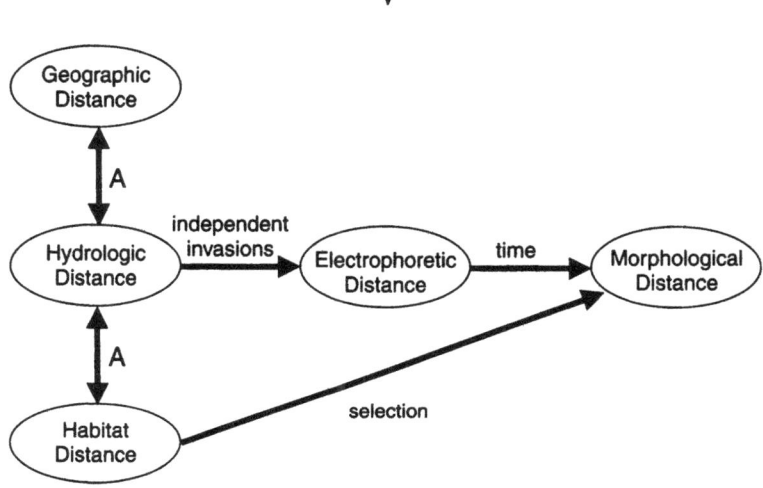

Fig. 7.3. Factors affecting morphological evolution in *Gammarus minus*. Only significant paths are shown (compare with Fig. 7.1).

is habitat distance that reflects the structure of the selective environment (*sensu* Brandon 1990). The strong dependence of morphological distance on habitat distance (Fig. 7.2) could reflect selection (path C of Fig. 7.1), purely environmental effects, or gene-environment interaction effects (path D of Fig. 7.1). Finally, the smaller but statistically significant relationship between electrophoretic distance and morphologic distance after the effects of habitat distance are removed indicates that morphological variation is also affected by the evolutionary history of the populations. We have summarized what we know about the evolution of *G. minus* in Figure 7.3.

Invasion of Caves and Isolation in Caves

The standard paradigm of cave invasion in temperate regions, dating back at least to the French cave biologist Rene Jeannel, a staunch neo-Lamarckian, is that caves served as refugia for ancestors of cave-limited species (troglobites) from the vicissitudes of Pleistocene climatic changes (Jeannel 1923). Later expanded and clarified in a neo-

Darwinian context by Barr (1968) and by Peck (1980), this view came to include the idea that evolution in caves corresponded exactly to Mayr's view of speciation via peripheral isolates. In brief, during periods of climatic warming, the cold-adapted fauna both of the forest floor (for example, trechine carabid beetles) and of small streams (for example, amphipods and isopods) colonized cave habitats when a few individuals of such species took refuge underground from climatic changes at the surface. These changes resulted in the extinction of the surface populations, and the subterranean populations became isolated.

Populations thus isolated were thought to be very small, and they therefore experienced a severe founder effect and, later, a "genetic revolution" resulting in the acquisition of troglomorphic features (Barr 1968). Improved adaptation to cave habitats in these forms presumably permitted extensive subterranean dispersal into more caves, limited only by the geographic extent and continuity of the limestone. For example, Barr (1979) contended that *Neaphaenops tellkampfi*, a troglobitic and highly troglomorphic carabid beetle found in scores of caves in west central Kentucky, invaded caves only once. Its current geographic range, the largest of any known cave beetle and including differentiation into four subspecies (Barr 1979, see also Kane and Brunner 1986), is assumed to be the consequence of spread via underground routes subsequent to cave isolation.

The Pleistocene refugium model contains both vicariant and dispersalist elements. These include:

1. Isolation of ancestral populations, without active invasion of caves
2. Genetic bottleneck followed by a reorganization of the epigenotype—Barr's (1968) "genetic revolution"
3. Subsequent subterranean dispersal and increase in geographic range as cave adaptation improves

In the case of some terrestrial species, intermediate stages of forms living deep in the soil have been proposed (Barr 1985), but even in these cases the scenario is one of the restriction of the ranges and the amount of gene flow among pre-cave ancestors (vicariance), followed by subterranean migration by incipient troglobites (dispersal). We like to think of this refugium model as vicariance in the small (isolation of populations) and dispersal in the large (Fig. 7.4).

Genetic data are equivocal in support of this hypothesis. Some cave-limited species show extremely reduced genetic variability (Laing, Carmody, and Peck 1976a,b, Kane, Barr, and Badaracca 1992), which is consistent with the occurrence of a genetic bottleneck, whereas other species (Crouau-Roy 1988, Kane, Barr, and Badaracca 1992) exhibit levels of variability that are not different from those of abundant, widely distributed surface-dwelling species. Likewise, although genetic data are sometimes consistent with the hypothesis of a single isolation in a cave of the surface ancestral population followed by subterranean migration (dispersalism), patterns of morphological and genetic differentiation in other troglobites may be equally well explained as a consequence of multiple isolations of the surface ancestors with subsequent convergent or parallel evolution among the descendants (see Laing, Carmody, and Peck 1976a,

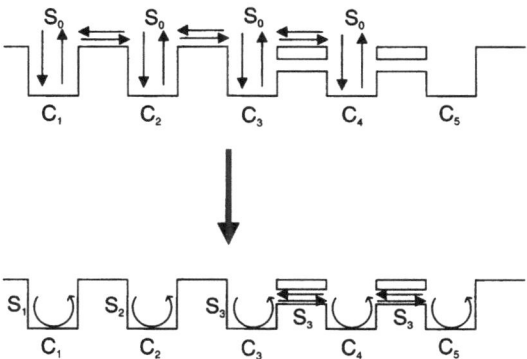

Fig. 7.4. Barr's model of evolution in caves. The top panel indicates gene flow *(arrows)* in a troglophilic species, S_0, found both in caves (labeled C_1 to C_5) and on the surface. Caves C_3, C_4, and C_5 are connected by subterranean channels. In the bottom panel surface populations have gone extinct, and populations S_1, S_2, and S_3 are descendents of S_0. Cave C_5, originally unoccupied by S_0, is subsequently colonized by S_3 via subterranean dispersal. Modified from Barr (1967).

Swofford 1982, Kane and Brunner 1986, Kane, Barr, and Badaracca 1992). The lack of extant surface populations in these cases is a severe impediment to our ability to distinguish among alternative hypotheses.

Recently, the Pleistocene refugium model has been challenged on several other fronts. It has not proven useful as an explanation for the origin of tropical troglobites, for which alternative models have been proposed. Howarth (1980) contends that many tropical cave species, including those of Hawaiian lava-tube caves, are derived from species that actively invaded caves in order to exploit a new habitat relatively rich in resources, root exudates in particular. Rouch and Danielopol (1987) argue even more generally that invasion of caves and other subterranean habitats is often an active process involving large rather than small populations.

Isolation is a key part of the refugium paradigm. Its importance is reflected in the very classification of cave organisms into troglobites (obligate cave dwellers), troglophiles (facultative cave dwellers), and trogloxenes (cave visitors) and in the polemic surrounding whether morphologically unmodified species known only from caves are troglophiles related to as yet undiscovered surface-dwelling populations or are troglobites (Hamilton-Smith 1971). Cave biologists have been strong supporters of models of allopatric speciation (see Sbordoni 1982, Barr and Holsinger 1985). The mere fact that there is a clear boundary between cave and surface habitats begs, almost from necessity, for a speciation model that involves geographic separation. However, there have been some recent challenges to the primacy of isolation from Wilkens and Huppop (1986) and Howarth (1993). They argue that adaptation and evolution can occur with contiguous cave- and surface-dwelling populations. Howarth suggests that cave populations can swamp surface populations. However, since troglobites, by definition, have no extant surface-dwelling populations, they are of little empirical use in evaluating the efficacy of parapatric and/or sympatric models of cave speciation.

In the following paragraphs, we describe in some detail what we know or can infer about the invasion of caves by *Gammarus minus* and the subsequent isolation of these populations. We do not claim that the species is typical of all those invading subterranean habitats or even typical of most. Indeed, our examination of this voluminous literature suggests there are no simple generalizations. We do claim that *G. minus* is typical of populations invading new habitats in general. Finally, we leave it to others to pigeonhole this particular case into the growing number of categories of speciation mechanisms.

What precipitated movement by *G. minus* into caves is a question about which we can only speculate. Present-day populations of *G. minus* in springs are sensitive to high temperature (see Chapter 4), and increases in summer temperatures could be avoided by invasion of caves. With summer temperture increases, the physical extent of the cold-water surface habitat shrinks and springs effectively become tiny cold-water islands connected to a cold-water subsurface continent. In addition, caves represent a predator-free environment. There are of course several immediate difficulties faced by a *G. minus* population invading a cave—absence of light and reduced food in particular.

The present-day distribution of *G. minus* in caves suggests that invasion of caves is not uncommon. The presence of *G. minus* in caves is in a state of dynamic flux. We have observed at least one invasion and one extinction of *G. minus*. Court Street Cave in Greenbrier County, West Virginia, in the middle of basin III (see Fig. 5.7) is a small cave that used to have large populations of the isopod *Caecidotea holsingeri* and the amphipods *Stygobromus emarginatus* and *S. spinatus* (Holsinger, Baroody, and Culver 1976). In 1978 the small sinkhole containing the entrance was enlarged to serve as a drain for an adjoining field. Within eighteen months of this change, *G. minus* had invaded the cave. Apparently, enlarging the sinkhole resulted in an increase in food in the stream, allowing *G. minus* to invade. This suggests that *G. minus* can rapidly take advantage of a suitable new habitat. We have also recorded one extinction. Hubricht (1943) reported a population of *G. minus* from Sinks of Gandy, a cave in northern West Virginia. Repeated visits by us in the 1980s and 1990s failed to relocate the population, which apparently has gone extinct.

Invasion of the large caves where troglomorphic populations of *G. minus* occur may be very infrequent, since the habitable spring is separated from the habitable cave stream by an apparently uninhabitable stretch of closed conduit. However, indirect evidence suggests that the invasion of these caves did not involve a small number of organisms with a resulting genetic bottleneck. Of all those studied in West Virginia, the population in the Bone-Norman Cave basin (basin I of Fig. 5.7) is the least differentiated morphologically (Table 5.5) and genetically (Fig. 5.10), and it presumably represents a recent invasion. It has levels of heterozygosity that do not differ from those of typical spring populations of *G. minus* or of freshwater invertebrates in general. The expected (Hardy-Weinberg) heterozygosity in this population is 0.100 and the observed heterozygosity is 0.087. The pattern of heterozygosity observed in the Norman Cave population (I3C of Fig. 5.7) is typical for troglomorphic populations in both Virginia and West Virginia.

For troglomorphic populations there is considerable isolation from resurgence populations, as indicated by the F-statistics discussed in Chapter 5. For all basins except basin I, Nm, the average number of migrants per generation, is less than 0.5. Likewise, the average number of migrants per generation between basins is less than 0.5. Morphological data bear this out as well. Spring phenotypes (representing upstream movement) have not been observed in over 5,000 individuals taken from caves. Cave phenotypes (representing downstream movement) have been observed less than 0.5 percent of the time in over 5,000 individuals taken from resurgences. Physically and ecologically, the populations are isolated.

While it has proven agonizingly difficult to breed *G. minus* in the laboratory, we have some information on reproductive barriers among populations. In forced hybridizations between cave populations of different basins, there was no reduction in the number of offspring (Table 7.4). Although mate-choice experiments have not been done, it is clear that individuals from cave populations in different basins will readily go into amplexus. This may be the least ecologically relevant comparison, since these populations are unlikely to

Table 7.4. Summary of preliminary hybridization experiment between cave-dwelling *G. minus* from two different basins (III and VI in Fig. 5.7). The two populations are Benedicts Cave (III4C) and Organ Cave—Organ Stream (VI13C). Values shown are mean number of offspring per female ± standard error. N = Sample size. Tukey's *a posteriori* w test indicated significant differences ($p < 0.01$) between the number of offspring produced by Organ Cave and Benedicts Cave females irrespective of the source of the father.

	Female source	
Male source	Organ Cave	Benedicts Cave
Organ Cave	5.67 ± 0.96 ($N = 9$)	12.11 ± 0.94 ($N = 9$)
Benedicts Cave	5.75 ± 1.10 ($N = 8$)	9.22 ± 1.47 ($N = 9$)

mix in nature except in the case of subterranean stream capture, such as has apparently happened between basins II and III. In matings between individuals from resurgence and cave populations there is a barrier. Large cave males do not always recognize the smaller resurgence females as a potential mate but rather may regard them as potential food. *G. minus* is not only a predator of other amphipods and isopods (Chapter 4) but a cannibal as well. In a series of mating experiments 8 of 25 female *G. minus* (32 percent) from Organ Spring put in a small aquarium with males from either Organ Cave or The Hole were eaten after one week. By contrast only 4 of 75 females (5.3 percent) were eaten when put in a small aquarium with males from the same population. In nature, it is likely that physical isolation and ecological isolation prevent gene exchange between cave and resurgence populations.

Gammarus minus has invaded karst basins at various times (see Chapter 8). Some of these invasions, as in the Sinks of Gandy, are ultimately unsuccessful. Some invasions never become isolated from surface ancestors, in large part because of the details of hydrogeology (see especially Figs. 5.3 and 5.4). Other invasions have been successful and the populations have become isolated from surface ancestors. The cause of the invasions may have been climatic, biotic, or a combination of both, but in any case large numbers of animals were involved. In contrast to the refugium paradigm of vicariance in the small and dispersal in the large, the invasion of caves by *G. minus* can be characterized as vicariance in the large and dispersal in the small. Vicariance is involved in the splitting of the initial populations; dispersal is involved in the invasion of caves within a basin.

Evolution of Troglomorphy

The cave environment of *G. minus* differs from the spring environment of *G. minus* in two fundamental ways: the complete absence of light and reduced levels of available food. In *G. minus* the absence of light results in strong directional selection for an increase in the relative size and complexity of extra-optic sensory structures. The complexity of extra-optic sensory structures in *G. minus* has been little studied. There are more antennal calceoli in troglomorphic individuals than in others (V. Steele, pers. comm.) but this may be due to size differences. The organ of Bellonci, an extra-optic sensory organ in *Gammarus* (Steele 1984) that is likely to show spring-cave differences, has not been studied in *Gammarus minus*. We did document strong directional selection for increased relative antennal size, overall size (which increases absolute antennal size), and reduced eye size in cave populations. Directional selection for reduced eye size is likely a reflection of events at the neurological level. Increased neurological connection to the brain of antennae may come only at the expense of decreased neurological connection to the brain of eyes. This is where the evolutionary tradeoff occurs. Strong selection coupled with high trait heritability and high trait variability leads to the evolution of troglomorphy.

While there was clear evidence for selection and clear evidence for differentiation of eye and antennal features between cave and resurgence populations, no such clear evidence exists for ability to survive in a food-poor environment. It is quite true that an increase in the size of extra-optic sensory structures should lead to an increase in food-finding ability, the increase should also lead to an increase in mate-finding ability, predator avoidance, and habitat location. On the other hand, there was no evidence of decreased metabolic rate in cave populations, perhaps because there was no genetic variation upon which selection could act or because increased food-finding ability, via elaborated extra-optic sensory structures, "solved" the problem of scarce resources.

One of the unsolved puzzles of the morphology of *G. minus* is the considerable variability within resurgence populations. As far as we can determine, *G. minus* and its antecedents have had a very long evolutionary history in cold-water habitats of springs and spring runs. Normally, one would expect selection to be largely in the form of stabilizing selection. Yet stabilizing selection is rare in *G. minus* and both eye and antennal morphology are quite variable.

Previously, we have referred to the independent acquisition of troglomorphic features by *G. minus* in different basins as an instance of parallelism. But is it really parallelism, or is it convergence? The difference between the two hinges on whether *G. minus* in different basins evolved from the same or different ancestors (Fig. 7.5). For example, there can be little dispute that the evolution of increased lateral line systems and eye loss in unrelated fish taxa that have invaded caves (Amblyopsidae in North America and *Caecobarbus* in Africa) is a case of convergence. The question is murkier the more closely related the groups are. Thus it is not clear whether the troglomorphy of the amblyopsid fish genera *Typhlichthys* and *Amblyopsis* is an example of parallelism or convergence. We believe that, in the case of the evolution of troglomorphy in *G. minus*, the available evidence points toward convergence rather than parallelism.

We will argue in Chapter 8 that *G. minus* invaded caves at least two and probably three different times separated by hundreds of thousands of years. In a temporal sense at least, they do not have the same common ancestor. Even in those instances of more or less simultaneous invasion of caves, the resurgence populations were likely to have been more or less distinct. If invasion occurred at a time of warming, especially in summer, then the distribution of surface-dwelling populations would be even more restricted than at present. Movement between springs would have been restricted and different

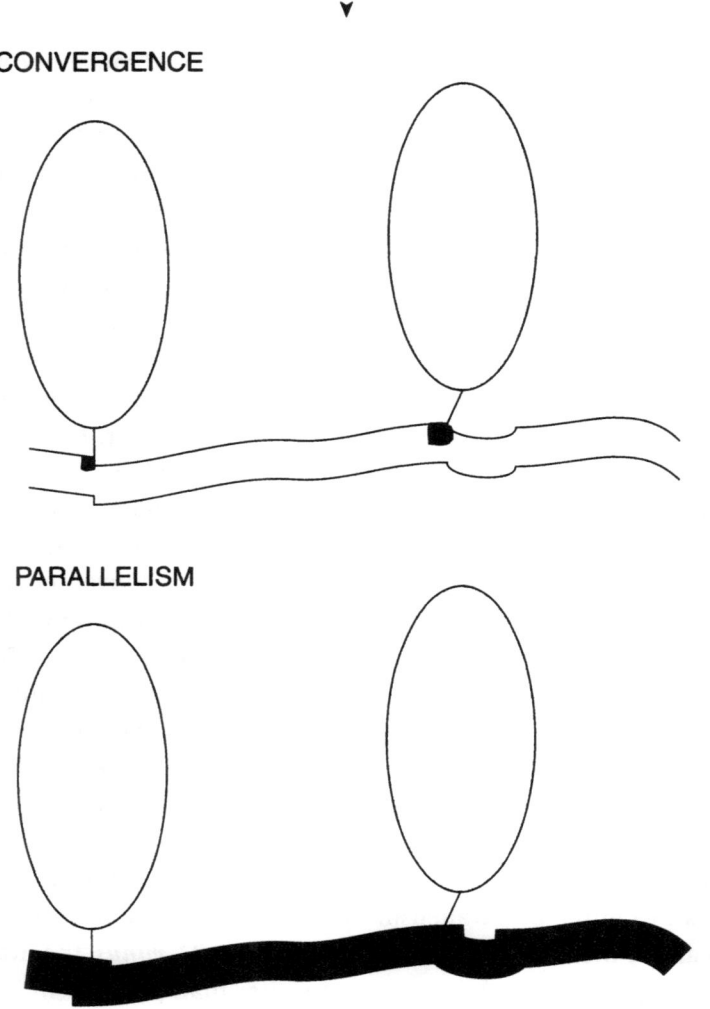

Fig. 7.5. A model of the difference between convergence and parallelism in the evolution of *G. minus*. The shaded ellipses represent subterranean basins that resurge at a stream via closed conduits. The initial distribution of *G. minus* in the stream is shown in black. If the convergence model is correct, *G. minus* moved into the two basins from two separate resurgence populations. If the parallelism model is correct, the basins were invaded by the same resurgence population. Available evidence supports the convergence model.

resurgence populations would have been more isolated one from the other. Even though current resurgence populations are very similar electrophoretically, other techniques detect differences. The biplot analysis (Jernigan et al. 1994) of genetic correlations showed a clear connection between resurgence and associated upstream cave population (Figs. 6.4 and 6.5). Preliminary analysis of mtDNA sequences of 16S rDNA genes in nine West Virginia populations also indicates differentiation among spring populations and some connections between resurgence and associated upstream cave populations. In particular Organ Cave resurgence (VI12R), nearby Dickson Spring (V11R), and Organ Cave–Big Canyon (VI18C, see Fig. 5.7) form a monophyletic group separated from basins II and III and their resurgences. If indeed *G. minus* in caves has undergone convergence, then what should be the taxonomic status of these populations?

How Many Species of *Gammarus minus* Are There?

We have approached the study of the evolution and adaptation of *G. minus* in caves from a microevolutionary point of view. That is, we have focused on changes through time of a lineage (anagenesis) rather than changes that might be observed resulting from the splitting of lineages (cladogenesis). Yet it is clear that isolation has been an important factor in the evolution of *G. minus*. Recall that troglomorphic populations occur only in large basins with little surface input and that resurgences and upstream caves are separated by a barrier, the phreatic loop (see Chapter 5). In the last fifty years only one paper has been published that attempted to address taxonomic problems involved with cave populations of *G. minus*, that of Holsinger and Culver (1970). There has been no extensive study of reproductive barriers and no search for unique, derived characters (synapomorphies) for cave populations in different basins. However, several points do emerge from the work that has been done.

First, species concepts involving reproductive isolation (or its inverse) do not seem central to the evolution of *G. minus* populations or their likely evolutionary fate. Once a *G. minus* population invades a subterranean basin, it is unlikely to encounter other *G. minus* populations. If *G. minus* invades a previously occupied basin, we suspect the animals are quickly eaten by the residents, especially since the residents are likely to be larger. It is unlikely to encounter populations from other basins except in the case of stream capture, as apparently happened in the northern part of the West Virginia study area.

Most important, species concepts that rely on either interbreeding (recognition species concepts) or failure to interbreed (isolation species concepts) do not take into account the major adaptive shift from springs to caves. Of course, some systematists would argue that such an adaptive shift is irrelevant to taxonomy, but any scheme that includes all cave and spring populations within one species no longer incorporates the idea of an "evolutionary species," one that shares a common evolutionary fate (see Templeton 1989). Christiansen and Culver (1968) made these same points in their study of the collembolan *Pseudosinella hirsuta*, which has invaded caves at different places and at different times. As with *G. minus*, reproductive isolation seems unimportant.

It is equally clear that there are not two species—one dwelling in springs and some caves and a second troglomorphic species in some caves (the variety *tenuipes* of Shoemaker 1940). *G. minus* has invaded caves repeatedly, and the isolated populations show strong convergent evolution. A species called *Gammarus tenuipes* would be polyphyletic, a characteristic that all the warring schools of taxonomy agree is undesirable.

The most reasonable approach to the taxonomy of *G. minus* seems to us to be a phylogenetic species concept. However, as Templeton (1989) points out, the concept of a shared common evolutionary fate is not an operational definition. We need to look for synapomorphies. At the genetic level we have indeed found some. In Ward Cove in Virginia, all troglomorphic individuals studied share a unique allele at the SOD-2 locus, while individuals from the resurgence populations share a different allele (Table 5.1). Other examples can be found in the West Virginia populations, but not all basins can be so easily defined. This is especially true for Davis Spring basin (III in Fig. 5.7), in which different branches may have been invaded separately and which has pirated part of the subterranean drainage of basin II.

There is a general reluctance to describe species on the basis of electrophoretic variation. It is as though taxonomy is really the domain of morphology, in spite of all the careful elaboration of species concepts over the past decades. It is certainly true that the description of "electrophoretic" species would lead to a proliferation of species. This may be inconvenient, but it would not necessarily be incorrect. There are indeed considerable problems with the description of cave-limited species based primarily on morphology. The major problem is that of parallelism and convergence. The problem with the genera erected by Lamarckian and neo-Lamarckian taxonomists in the late 1800s was that they were polyphyletic. This same problem can reappear in a more modern guise as more and more cave taxa are subjected to cladistic analysis. In practice, most synapomorphies used to separate subterranean taxa are character reductions, and, in the terminology of Christiansen (see Chapter 2), they are likely to be cave-dependent characters.

In one of the most thoughtful and thorough such studies to date, Boutin, Messouli, and Coineau (1992) did a cladistic analysis of a group of obligate subterranean amphipods (Metacrangonyctidae) from Morocco. Of the twenty-two characters used in the phylogenetic analysis, the apomorphic state of seventeen was either absence or reduction. Examples of apomorphic characters are the absence of the endopodite of uropod 3 and a reduction in teeth of the lacinia mobilis. The adaptive significance of such characters, if any, is unknown. It comes as no surprise that the primary focus of this paper is an examination of the efficacy of different parsimony programs in detecting homoplasy in this data set. In fact, for *G. minus* in several subterranean basins, cave-dependent characters cannot be used to define taxa because of the presence of karst-window populations, which are exposed to sunlight and which show varying degrees of the apparent re-acquisition of large eyes and small antennae.

Good taxonomy of cave animals always has a strong geographic and hydrogeologic component. This is true of the work of Boutin and his colleagues and it is true in North America of the amphipod taxonomy of Holsinger. The explicit or implicit yardstick of comparison for various taxonomic concepts when applied to cave organisms is the geographic yardstick.

One of the consequences of this geographical yardstick is that when surface ancestors are extinct, many species are described (for example, more than 200 species of *Pseudanophthalmus* beetles), and few are described if surface ancestors are extant, even if there is no gene flow. This is true not only of *G. minus* but also of the collembolan *Pseudosinella hirsuta* (Christiansen and Culver 1968), hydrobiid snails (Hershler, Holsinger, and Hubricht 1990), *Asellus aquaticus* (Sket 1965), and the Mexican cave fish *Astyanax fasciatus* (Mitchell, Russell, and Elliott 1977). As a conservative taxonomic practice, there is much to applaud in the dichotomy between groups with and without extant closely related surface populations, but it may also serve to obscure the connection between these polytypic taxa and the highly restricted (both geographically and morphologically) taxa.

Variation in *Gammarus minus* and other such species is not merely intraspecific variation. It is variation that, when surface populations go extinct, leads to the description of highly speciose groups. For cave populations of *G. minus* and similar species, the presence or absence of surface populations is irrelevant to its ecological and evolutionary dynamics, in both the short and the long term.

Summary

We analyzed five different measures of distance among cave, karstwindow, and spring populations. Three of them—geographic distance, hydrologic distance, and habitat distance—reflect physical distance among populations. The remaining two—electrophoretic distance and morphological distance—reflect biological distance among populations.

Putting the Pieces Together

▼

We explored the roles of evolutionary history and of natural selection in the evolution of *G. minus* by examining causal paths relating the physical variables to the biological variables reflected in the matrices. Subsurface connectivity among populations as measured by hydrologic distance is the most significant variable for understanding the genetic structure of the populations as reflected by electrophoretic distance. We interpret the causal path linking hydrologic and electrophoretic distances as that of invasion history, especially of independent invasion into different subterranean drainage basins.

The significant correlation between electrophoretic and morphological distances also suggests that morphological variation among populations is influenced by evolutionary history. There is no significant path linking habitat and electrophoretic distances, indicating that selection has little influence on genetic structure as measured by allozyme variation. The significant path linking habitat and morphological distances does indicate that selection has a strong effect on morphological variation among populations.

Invasion of caves by *G. minus* appears to be an active process, most likely because caves present a relatively predator-free environment with a larger extent of a suitable thermal regime. Whereas such invasions are common, successful establishment of cave populations isolated from surface ancestors is less common. Cave populations are subject to strong directional selection for increased relative antennal size and reduced eye size. However, extant troglomorphic populations retain substantial morphological variation. The independent acquisition of troglomorphy by *G. minus* in different basins appears to be a case of convergence. We reason that ancestral populations of the different troglomorphic cave populations are temporally separated by hundreds of thousands of years, or are spatially separated by isolation in springs, effectively thermal islands.

This analysis leads to the conclusion that what is known as *Gammarus minus* is actually a species complex. Populations in resurgence habitats constitute one species. Further, genetic and morphological differences indicate that there may be as many troglomorphic species are there are independent isolations in different basins.

Selected References

Barr, T. C. 1985. Pattern and process in speciation of trechine beetles in eastern North America (Coleoptera: Carabidae: Trechinae). In *Taxonomy, phylogeny, and zoogeography of beetles and ants*, ed. G. E. Ball, pp. 350–407. Leiden: W. Junk. Includes an updated view of isolation and speciation of cave carabid beetles.

Barr, T. C., and J. R. Holsinger. 1985. Speciation in cave faunas. *Annual Review of Ecology and Systematics* 16:313–337. A review of modes of speciation in cave organisms.

Boutin, C., M. Messouli, and N. Coineau. 1992. Phylogénie et biogéographie évolutive d'un groupe de Metacrangonyctidae, Crustacés Amphipodes stygobies du Maroc. II. Cladistique et paléobiogéographie. Avec l'examen comparatif de plusieurs logiciels de parcimonie. *Stygologia* 7:159–178. A thorough cladistic analysis of an exclusively subterranean group.

Manly, B. F. J. 1991. *Randomization and Monte Carlo methods in biology*. London: Chapman and Hall. Techniques for matrix regression as well as relevant examples.

Rouch, R., and D. Danielopol. 1987. L'origine de la faune aquatique souterraine, entre le paradigme du refuge et le modèle de la colonisation active. *Stygologia* 3:345–372. A challenge to the classic view of passive isolation of populations in caves.

8 Questions of Time
▼

We have thus far paid only passing attention to an issue that was and is of considerable importance to understanding the evolution of cave organisms—that of time. Neo-Lamarckians insisted on a rapid evolution of troglomorphy (see Chapter 2). Neo-Darwinians such as Barr (1968), on the other hand, dismissed neutral mutation and genetic drift as a cause of eye and pigment loss because they held, wrongly it turns out (Culver 1982, Nei 1987), that neutral mutation proceeds too slowly to account for the observed morphological change. Time, in a relative sense, also figures prominently in the work of the neo-Darwinian Poulson (1963) and the neutralist Wilkens (1973, 1986). Both use the relative amount of eye and pigment loss in cave fish to order a temporal series. Wilkens (1986) correlates different stages of morphological change with periods of marine inundations in Yucatán and other parts of Mexico, but there is no direct evidence for this correlation. In a similar way, Sbordoni and colleagues (1980) use heterozygosity levels of electromorphs to estimate time elapsed since the apparent stranding of isopods from predominantly marine groups in caves.

In this chapter we take up the issue of time as it concerns *Gammarus minus*. Four interconnected questions are of interest here. The first is the age of the caves and the age of the subsurface basins. Most cave biologists have ignored this question, although it has occupied a great deal of attention on the part of karst geomorphologists. To the cave biologist caves and the karst landscape may appear practically eternal, but this need not be the case. For example, significant karst development has taken place in San Salvador, Bahamas, within the past 10,000 years (Mylroie and Carew 1986). We examine estimates of the age of the caves and karst in the areas where troglomorphic populations of *G. minus* have evolved.

Second, we must consider how long ago caves were invaded by *G. minus*. We obtain estimates from the allozyme data previously discussed in Chapter 5 and from some preliminary data on sequence divergence of a mitochondrial gene. Then, given estimates of the age of invasion, we come to the third question: is the rate of morphological change commensurate with the amount of change expected under neutral mutation and genetic drift? This question effectively turns the problem of the role of selection on its head. That is, is selection, either stabilizing or directional, required to explain the observed rate of morphological change?

The fourth question involves the reversal of morphological change. Karst-window populations, with variable but intermediate morphology between resurgence and cave populations, appear to represent a reversal of the evolution of troglomorphy. Have they reevolved the eye and pigment? Using a combination of morphological, allozyme, and RAPD (random amplification of polymorphic DNA) data, we show that traits that have been lost in troglomorphic populations do indeed reappear in karst-window populations.

Age of Caves and Karst

Trying to determine the age of a cave is, in effect, trying to determine the age of a void. Several approaches to the problem have been tried, including uranium series dating, electron spin resonance, thermoluminescence, magnetic records, and carbonate denudation rates (White 1988). Of these, only magnetic records and carbonate denudation rates may be used for materials more than a million years old.

The most thorough study of paleomagnetic records of cave sediments is that of Schmidt (1982) for Mammoth Cave, Kentucky. Mammoth Cave is unusual in that old passages, rather than being eroded away, are preserved underneath a sandstone caprock. This in part explains the immense size of the cave, with well over 400 km of passages. Sediment in the highest (and oldest) passages are late Pliocene, approximately 1.7 to 2.1 million years of age. As White (1988) points out, individual conduits have a discrete life history (Fig. 8.1), as he terms it. The history begins with the slow dissolution of a fracture under laminar flow. After some 3,000 to 5,000 years, the fracture reaches a threshold diameter of about 10 mm, after which it rapidly enlarges. Eventually the conduit "dies," destroyed by collapse, fragmentation, and removal by surface erosion. According to White, destruction occurs after 10^6 to 10^8 years (Fig. 8.1), depending on the geologic setting.

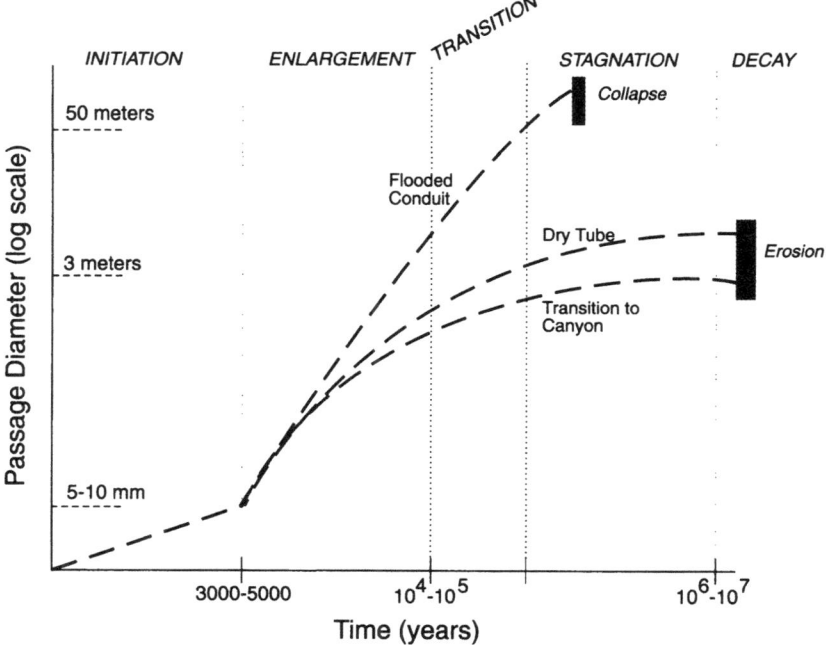

Fig. 8.1. Schematic rate curve for the "life history" of a single conduit. Black rectangles indicate "death" of conduit due either to collapse or fragmentation and removal by surface erosion. From White (1988).

It is not an individual conduit that *G. minus* invades, however, but rather a subterranean basin. White and White (1991, see also White 1988) attempt to answer the question of basin age using simple calculations of the denudation rates of karst in the Appalachians. In regions of non-carbonate rock, regions without caves, the landscape changes as downcutting valleys take on a characteristic V-shape, interspersed with broad floodplains of erosion formed during periods of base-level stability. In contrast, karst areas need not change their form; they are simply lowered at a rate determined by the rate of dissolution of the underlying carbonate rock (White and White 1991). After estimating the rate of downcutting, White and White estimate that it would take approximately 2 to 5 million years to downcut approximately 700 meters, the present difference between the surface of the karst basin and the elevation of surrounding non-carbonate rock. This estimate is based on a cyclical view of the evolution of landscape, introduced by W. M. Davis in 1889. That is, at the start (and end) of a cycle, a broad peneplain, or flat, eroded plain, existed, in this case the Harrisburg Peneplain. At this stage, there is unlikely to have been any suitable subsurface habitat for *G. minus*. If such neo-Davisian views as the Whites set forth are incorrect, and a model of continuous evolution of the landscape (Hack 1960) is correct, then karst basins may be much older. We can, however, take as a best estimate an age of 2–5 million years.

Age of Invasion and Isolation of *G. minus* in Caves

The fossil record provides only the most meager of clues about the age of separation of *G. minus* from other species of *Gammarus* and no information about the age of invasion of caves. While the fossil record of amphipods in general is disappointing, we do know that the genus *Gammarus*, which is well preserved in imprint fossils from the Oligocene (Barnard and Barnard 1983), is at least 2.5×10^7 years old. There is no fossil record for *G. minus* itself, and the time of its separation from other North American *Gammarus* species is a matter of speculation.

Genetic data can be used to provide dates for the isolation of *G. minus* in caves. Nei (1987) estimates that the average rate of codon substitution detectable by electrophoresis is 10^{-7} per locus per year. From this rate and from Nei's electrophoretic distance, D, we can make an estimate of time since isolation, t:

$$t = 5 \times 10^6 D \tag{8.1}$$

We used an electrophoretic distance between populations from caves and their resurgence to estimate t (Table 8.1). Estimates for the time since isolation range from 100,000 years for basin I (Bone-Norman) to 670,000 years for basin VI (Organ). The four basins with the greatest differentiation in eye morphology (basin VII in Virginia and basins II, III, and VI in West Virginia) were also isolated the longest time, between 300,000 and 670,000 years, according to equation (8.1). Estimates of time since isolation for the remaining basins, with

Table 8.1. Estimates of age (in years) of isolation of *G. minus* in caves in six West Virginia basins and one Virginia basin (Figs. 5.7 and 5.8). The first estimate (Time1) is based on the unbiased genetic distance for 18–19 loci of Nei (1987). The second estimate (Time2) is based on percent sequence divergence in 16S rDNA, a mitochondrial gene. The average distance between cave populations in a basin and those in its resurgence was used to find Nei's electrophoretic distance *(D)* and percent sequence difference (% Diff.). For basins II and III, the northern populations of basin III were considered to have originated in basin II and later captured by basin III (see Chapter 5).

Basin	D	Time1	%Diff.	Time2
I—Bone-Norman	0.020	1.0×10^5	—	—
II—The Hole	0.114	5.7×10^5	3.8	1.9×10^6
III—Davis Spring	0.059	3.0×10^5	3.1	1.5×10^6
IV—Scott Hollow	0.045	2.2×10^5	—	—
V—Dickson Spring	0.049	2.4×10^5	0.7	3.5×10^5
VI—Organ Cave	0.134	6.7×10^5	0.9	4.5×10^5
VII—Ward Cove	0.087	4.2×10^5	—	—

less differentiation in eye morphology (basins I, IV, and V in West Virginia) ranged between 240,000 years and 100,000 years. All of these estimates of time since isolation are considerably less than the age of the development of the karst basins, suggesting that invasion occurred in a landscape very much like the present one in basins in very much the same position as the present basins. The one proviso to this conclusion is that there apparently has been subterranean stream capture by basin III of streams in basin II (see Chapter 5).

A second, more preliminary estimate of age of isolation in subsurface basins is available (Kane in prep.). This is based on sequence differences in a portion of the 16S rDNA mitochondrial gene (Table 8.1). While no rate estimates of 16S rDNA in Crustacea are available, DeSalle et al. (1987) estimate the rate of nucleotide substitution at 10^{-8} per site per year on the basis of data from the cytochrome b mtDNA gene in *Drosophila*. The rate of divergence is then

$$t = 5 \times 10^5 d* \quad (8.2)$$

where $d*$ is the percent nucleotide difference. Estimates for the age of isolation for four basins are available (Table 8.1). For basin V (Dickson) and basin VI (Organ), there is good agreement between the mtDNA and protein estimates (Table 8.1). However, in the case of basin II (The Hole) and basin III (Davis Spring), which lie in the center of the Greenbrier Valley in West Virginia (Fig. 5.7), the time estimates based on mtDNA are nearly an order of magnitude larger than estimates based on electrophoretically detectable proteins (Table 8.1). Even these longer estimates are still within the bounds of the likely age of the present basins, roughly 2 million years (see above). One final estimate of age of separation is available from Kane's preliminary data. Using *Gammarus fasciatus* as the outgroup, he found the time since divergence of *G. minus* to be 10^7 years, during the mid-Miocene.

Taken together, these time estimates suggest multiple invasions of subsurface basins during the Pleistocene, the oldest perhaps 2 million years ago and the most recent perhaps 100,000 years ago. These molecular-clock estimates are at least not inconsistent with either the fossil record or estimates by geomorphologists of the age of the subsurface basins.

Time and Neutral Mutation

We have argued that natural selection is critical to an understanding of the morphological evolution of *G. minus*, including eye loss, but this is not to say that neutral mutation—mutation resulting in no selective advantage or disadvantage to an organism—plays no role. A main argument used to dismiss neutral mutation, however, has been insufficient time. For example, Barr (1968), citing the earlier work of Ludwig (1942), claims that neutral mutation and genetic drift would act too slowly to account for the observed patterns of regressive evolution. Indeed, this is likely to be true in the case of an infinitely large population where the structure in question is controlled by a single gene. Of course, cave populations are not infinite nor are eyes and pigment controlled by a single gene.

According to Nei (1987), the mean time, t, at which an allele is lost is

$$t = 4N_e + 1/v \qquad (8.3)$$

where N_e is the effective population size and v is the mutation rate, and the starting gene frequency, p, is 1. If we assume $v = 10^{-6}$ and $N_e = 10,000$, then the mean time to loss is 1.04×10^6, which is at the upper limit of our estimate of the time available for evolution of cave populations of *G. minus* (Table 8.1). In their discussion of eye loss in *Astyanax*, where time available for regressive evolution is likely even shorter, Nei and colleagues (Chakraborty and Nei 1974, Nei 1987) get around this problem by assuming a mutation rate to nonfunctional alleles of 10^{-5} per locus per generation, a rate an order of magnitude higher than would appear reasonable. As Prout (1964) pointed out, it is not necessary for any particular gene to be lost in a population, since many genes control structures such as eyes and the loss of any of these genes could result in eye loss. That many genes control eye development and that different genes are lost in different lineages is strongly suggested by Dai's (1989) data (Table 5.4) on *G. minus* and Wilkens's hybridization experiments with different cave populations of *Astyanax* (Wilkens 1971). Wilkens found that hybrids had larger eyes, suggesting that different genes were lost in different lineages. His data can be used to estimate the number of loci involved, which at a minimum is five (Culver 1982).

Consider a case where the loss of any one of five genes at independent loci results in eye loss. Nei (1987) shows that the probability that the original allele at a frequency of 1 becomes lost from the population by generation t is

$$f(t) = 1 - e^{-vt} \tag{8.4}$$

If time available is 5×10^5 years and $v = 10^{-6}$, then the probability a particular allele is lost is 0.40. However, the probability that none of the five alleles is lost is only 0.08 $[(1 - 0.4)^5]$. If more genes are involved or if the time available is greater, this latter probability becomes even smaller. If ten alleles are involved, the probability that none of the alleles is lost is 0.006. The point is not that neutral mutation has caused the observed eye reduction in *G. minus* or that five or ten loci are involved. The point is that given the length of time since isolation in caves, neutral mutation could account for the observed eye reduction in *G. minus*.

Further insight can be obtained by considering observed rates of morphological evolution and expected rates of morphological evolution resulting from neutral mutation. Lynch (1990) suggests the following measure of evolutionary rate, Δ, that is independent of measurement scale:

$$\Delta = \frac{var_B(\ln z)}{t\, var_W(\ln z)} \tag{8.5}$$

where $var_W(\ln z)$ and $var_B(\ln z)$ are the observed within- and between-population variance components for log-transformed measures of morphology, z. Although Lynch (1990) employs this measure for different species, we can apply the model since, as we have argued previously (Chapters 5 and 7), there is little or no gene flow between resurgence and subterranean basin populations. Using Fong's (1989) morphological measurements on populations from Organ Cave and Organ Cave resurgence, Δ was calculated for head length, two eye measurements, and six antennal measurements (Table 8.2). Assuming a divergence time of 5×10^5 years, Δ averaged 9.5×10^{-6} per year for antennal characters and 2.0×10^{-5} per year for eye characters, an order of magnitude greater. The range of Δ for *G. minus* is nearly the same as that for rates of morphological evolution of non-human mammals recorded by Lynch (Fig. 8.2).

Lynch (1990) also argued that the expected rate of morphological change under a neutral model would be between 5×10^{-5} and 5×10^{-3}. This range was calculated using estimates of mutation rates—actually estimates of variance due to mutation divided by environmental variance (Lynch 1988)—from a wide variety of animals, pri-

Table 8.2. Evolutionary rates of divergence, Δ (= $var_B(\ln z)/[t \times var_w(\ln z)]$), where $var_B(\ln z)$ and $var_w(\ln z)$ are between- and within-population phenotypic variance for log-transformed measures of morphology, z, for Organ resurgence (VI12R in Fig. 5.7) and Organ main stream (VI13C in Fig. 5.7) populations of *G. minus*. Time since divergence, t, was set to 5×10^5 years (see Table 8.1). Traits (see Table 6.1) are head length (hl), eye ommatidia number (en), surface area of compound eye (ea), peduncle length of first antenna (p1), flagellum length of first antenna (l1), number of flagellar segments of first antenna (n1), and the second-antenna analogs (p2, l2, and n2).

Morphological character		Δ	$var_B(\ln z)$	$var_w(\ln z)$
Size:	hl	4.0×10^{-7}	.00248	.01237
Eyes:	en	2.4×10^{-5}	.79983	.06601
	ea	1.5×10^{-5}	.42515	.05784
Antennae:	p1	8.2×10^{-7}	.00890	.02168
	l1	4.1×10^{-7}	.00544	.02677
	n1	1.1×10^{-6}	.01165	.02156
	p2	1.1×10^{-6}	.01315	.02472
	l2	8.9×10^{-7}	.01472	.03314
	n2	1.4×10^{-6}	.00996	.01450

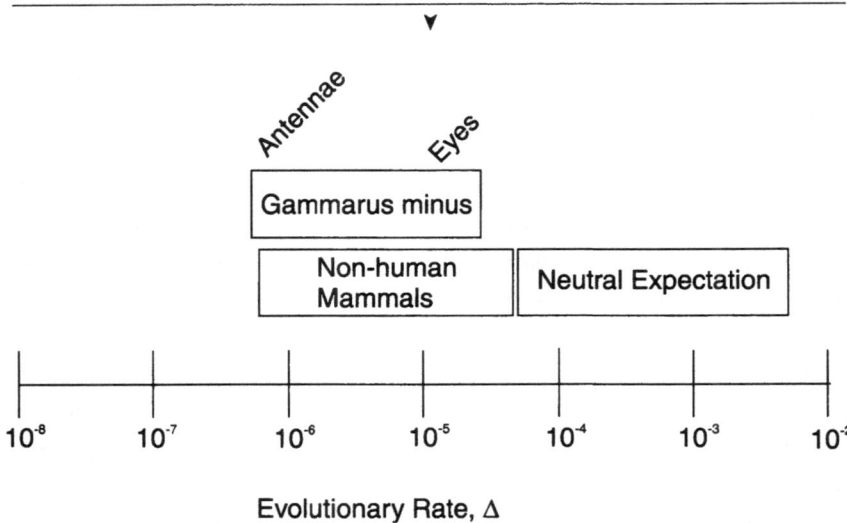

Fig. 8.2. Evolutionary rates, Δ, observed for *G. minus* (Table 8.2) and non-human mammals (Lynch 1990) and expected under neutral mutation (Lynch 1988, 1990). Rates are scaled per year.

marily vertebrates. Rates of evolution of both eye and antennal characters in *G. minus* are less than neutral expectation. This has also been found to be the case for rates of evolution of morphological characters in non-human mammals (Lynch 1990). If rates of evolution really are lower, rather than higher, than neutral expectation, stabilizing rather than directional selection is suggested. However, it is directional rather than stabilizing selection that predominates in *G. minus* (Chapter 6). In the absence of direct information about mutational variance in *G. minus*, the expected evolutionary rates under neutrality must be viewed with great caution.

The difference in rates between antennal and eye characters is particularly interesting. This difference in rates of change between elaborated and reduced characters is likely to be typical of other cave animals, such as *Astyanax* (Wilkens 1988). It may be the result of differences in mutation rates between eyes and antennae. That is, more genes may be involved in external eye morphology (size and ommatidia number) than in external antennal morphology (length and number of segments). However, it is interesting to speculate that it may also be due to the addition of neutral mutation to the rate of change of eye characters while antennal characters change largely as a result of selection. That is, there is likely to be a tradeoff at the neurological level between optic and extra-optic sensory structures. It is mutations acting at this phenotypic level that are acted on by selection. These mutations will also have a pleiotropic effect on characters we measured, such as size and length. On the other hand, some mutations will affect eye or antennal size without affecting neurological connections. These mutations will be neutral with respect to selection. Finally, reducing mutations are much more likely than elaborating mutations, as was first suggested by Fisher (1930). For eyes (and pigment), reducing mutation may be neutral and elaborating mutations selected against in the cave environment; for antennae, though, reducing mutation will be selected against in the cave environment.

Karst-Window Populations and Dollo's Law

Four populations in our West Virginia study area and one population in our Virginia study area are in what we termed karst windows—habitats well upstream of base-level resurgences that are exposed to light. All probably formed well after the karst basin developed. They are the result of cave passage collapse (Taylor Spring in basin I, Higginbothams Cave in basin III, and Hugh Young Cave in basin VII) or the intersection of a cave passage with a valley (Milligan Creek in basin III and Boyd Spring in basin V). Except for Milligan Creek, all are in close proximity to well-differentiated cave populations. Even though some are near enterable caves and some are not, the populations are always concentrated within a few meters of the exit of water to the surface. In addition, all karst windows are short in linear extent (less than 100 m) and populations experience fewer competitors or predators than are found in caves and resurgences.

The question that arises is whether karst-window populations are remnants of ancestral surface-dwelling populations or whether they arose from morphologically reduced cave populations. In terms of eye morphology, the karst-window populations range from having characters indistinguishable from those in a resurgence population (Taylor Spring) to having characters indistinguishable from those in a highly modified cave population (Hugh Young Cave). All except the Taylor Spring population have morphologically reduced eyes relative to resurgence populations (Fig. 8.3). With the exception of Milligan Creek, however, all populations are more similar electrophoretically to nearby cave populations than to resurgence populations. This is possibly the result of gene exchange beteen cave and karst-window populations. Two karst-window populations that we extensively studied are very different morphologically. The Hugh Young Cave population (VII26KW, Fig. 5.8) has greatly reduced eyes (mean ommatidia number = 0.8) and the Boyd Spring population (IV22KW, Fig. 5.7) has slightly reduced eyes (mean ommatidia number = 19.6).

RAPD (random amplification of polymorphic DNA) data were scored using presence/absence of bands as character states (Williams et al. 1990). Maximum parsimony trees were generated using Felsenstein's (1985) bootstrap technique in PAUP (Swofford 1991) for the Hugh Young Cave karst-window and associated populations in Virginia (basin VII, Fig. 5.8) and for Boyd Spring karst-window and associated populations in West Virginia (basin IV, Fig. 5.7). In spite of the morphological differences between the two karst-window populations (Fig. 8.3), the trees for these two basins do not reflect these morphological differences (Fig. 8.4). In both cases karst-window individuals cluster with cave individuals. In basin VII, cave and karst-window populations are a completely unresolved polytomy, and in basin IV, there is only partial separation of karst-window and cave populations (Fig. 8.4).

All of the genetic evidence, including allozyme and RAPD data, points to the conclusion that karst-window populations evolved from cave populations in the same basin. The evolutionary history of a karst-window population includes a period of eye reduction corresponding to time in caves followed by a period of eye increase corresponding to time in karst windows. They have thus re-acquired com-

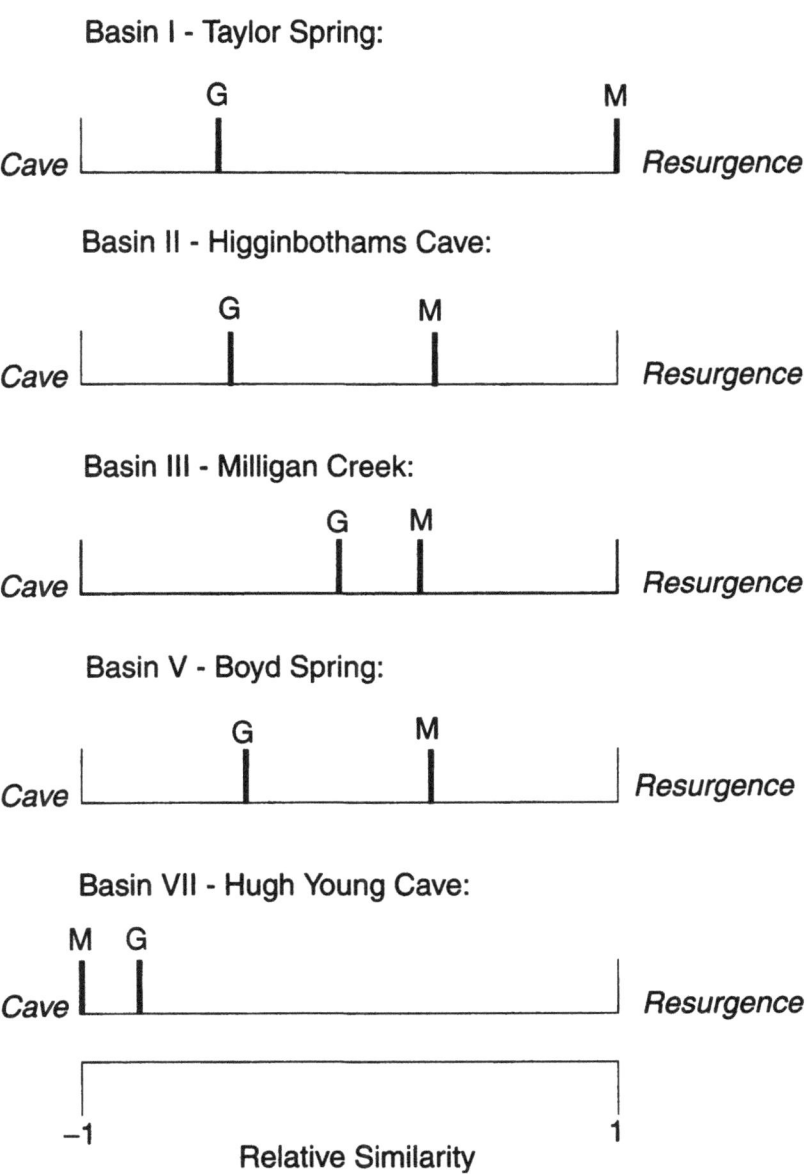

Fig. 8.3. Relative similarity of six karst-window populations, genetically (indicated by G) and morphologically (indicated by M), to their respective resurgence and cave populations in the same basin. Morphological similarity is based on eye ommatidia numbers (Table 5.5). Genetic (electrophoretic) similarity is based on Cavalli-Sforza and Edwards arc distances (see Fig. 5.10). The population of Higginbothams Cave in basin III is compared with basin II populations since it probably arose from there (see Chapter 5).

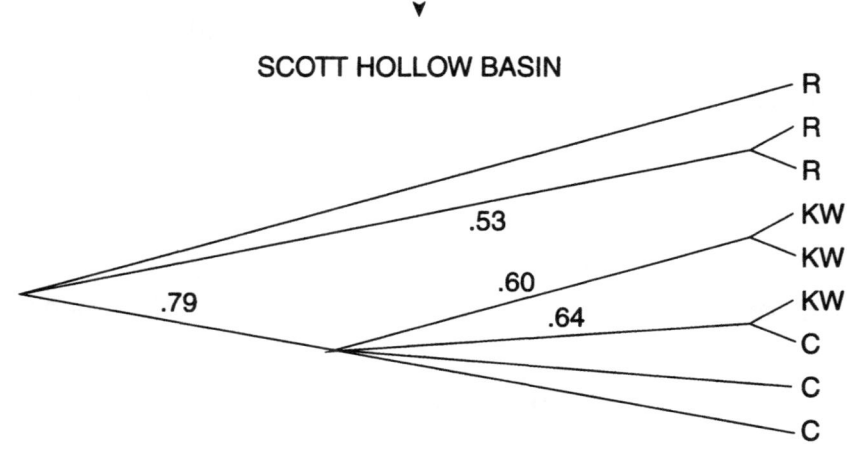

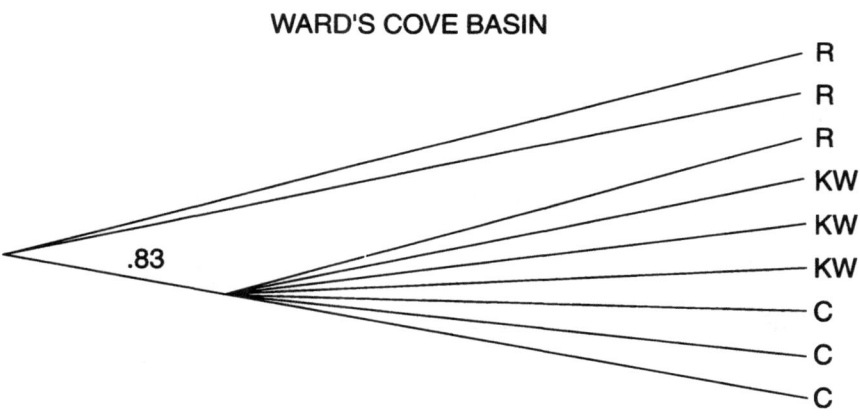

Fig. 8.4. Bootstrap majority rule consensus trees for *G. minus* populations of Scott Hollow basin (IV in Fig. 5.7) and Ward Cove basin (VII in Fig. 5.8). The tree for Scott Hollow was based on 29 DNA fragments, the one for Fallen Rock was based on 94 DNA fragments. Populations analyzed were from Scott Hollow resurgence (R), Boyd Spring (KW), and Scott Hollow Cave (C) in Scott Hollow basin; and Maiden Spring (R), Hugh Young Cave (KW), and Fallen Rock Cave (C) in Ward Cove basin. Numbers are bootstrap frequencies based on 500 resamplings (Felsenstein 1985).

pound eyes in apparent violation of Dollo's law of the irreversibility of evolution. While we have no way of estimating the age of colonization of karst windows, it is obviously less, and probably much less, than the age of karst basins. How can such a relatively rapid reversal occur? We suspect that cave populations of *G. minus* retain many of the genes for eye formation, but the genes are simply not expressed. This is the situation that Langecker, Schmale, and Wilkens (1993) found for the opsin gene in cave populations of the fish *Astyanax fasciatus*. No crystallin was present in the organisms but the crystallin gene was retained. By this hypothesis, in the reversal of eye reduction in karst-window populations, the genes are once again expressed.

Summary

The best estimate of the age of karst basins in the Appalachians is 2–5 million years. Estimates of time of invasion suggest that *G. minus* invaded underground basins multiple times during the Pleistocene, from as early as 2 million years ago to as recently as 100,000 years ago. Although we provide evidence for directional selection in the evolution of troglomorphy, these estimates suggest that, if our estimates of mutation rates and number of loci governing eye development are reasonable, neutral mutation could account for the observed degree of eye reduction. Observed rates of morphological change in *G. minus* are actually less than expected from neutral models. Eye size has evolved at a faster rate than antenna size. This may result from the two sets of characters having different mutation rates. More interestingly, it may reflect that while the evolution of antenna size is largely due to selection, the evolution of eye size is governed by additive effects of neutral mutation and selection.

Karst-window populations generally exhibit troglomorphic character states which are intermediate between those found in cave and resurgence populations in the same basin. The genetic evidence indicates that karst-window populations are derived from the cave populations. Apparently karst-window populations in different basins are strongly convergent in terms of reacquiring large compound eyes.

This suggests that eye reduction in the original cave populations may result from selection for degenerative mutations occurring at the regulatory loci governing eye expression while the structural loci remain relatively intact.

Selected References

Langecker, T. G., H. Schmale, and H. Wilkens. 1993. Transcription of the opsin gene in degenerate eyes of cave-dwelling *Astyanax fasciatus* (Teleostei, Characidae) and of its conspecific epigean ancestor during early ontogeny. *Cell and Tissue Research* 273:183–192. A seminal paper in the debate over whether genes have disappeared in regressive evolution or whether they are not transcribed.

Lynch, M. 1990. The rate of morphological evolution in mammals from the standpoint of neutral expectation. *American Naturalist* 136:727–741. Exposition of neutral models for morphological evolution.

Nei, M. 1987. *Molecular evolutionary genetics*. New York: Columbia University Press. The standard reference for determining the time scale for molecular change.

White, W. B. 1988. *Geomorphology and hydrology of carbonate terrains*. New York: Oxford University Press. A detailed description of karst basin and cave "evolution."

9 Adaptation in
▼ *Gammarus minus*

Levins and Lewontin (1985) point out that any theory that purports to generality contains the seeds of self-caricature. The theory of adaptation is certainly no exception. There is an adaptationist explanation, albeit extremely convoluted ones in some cases, for the evolution of any structure or any behavior. To avoid such self-caricature Brandon (1990, p. 165) has suggested five requirements for a "complete adaptation explanation." These are:

1. Evidence that selection has occurred, that is, that some types are better adapted than others in the relevant selective environment (and that this has resulted in differential reproduction)
2. An ecological explanation of the fact that some types are better adapted than others
3. Evidence that the traits in question are heritable
4. Information about the structure of the population from both a genetic and a selective point of view, that is, information about patterns of gene flow and patterns of selective environments
5. Phylogenetic information concerning what has evolved from what, that is, which character states are ancestral and which are derived

Others (see Levins and Lewontin 1985 and Sober 1984) have suggested very similar requirements but Brandon's are perhaps the most stringent. In this chapter we review the evidence that the morphology of *G. minus* is adapted to the cave environment and show how Brandon's criteria also serve to highlight the advantages of using *Gammarus minus* as a model in the study of adaptation.

Evidence That Selection Has Occurred

If adaptation is a consequence of evolution by natural selection, selection is a prerequisite for adaptation. We have presented two bodies of evidence that support the hypothesis that directional selection for troglomorphic characters has occurred. The first is the direct measurement of selection on cave and spring populations using multivariate regression techniques. In particular, directional selection for increased antennal size and decreased eye size occurred in all three cave populations studied (Chapter 5, and Jones, Culver, and Kane 1992). The shape of the selection curve in cave populations was predominantly monotonic (Chapter 5, and Culver et al. 1994). The second body of evidence is the phenomenon of troglomorphy itself. For cave animals in general and *G. minus* in particular, series of parallel and convergent morphologies are observed, especially the elaboration of extra-optic sensory structure and the reduction of optic structures. Independently derived cave populations of *Gammarus minus* (see Chapter 7) show increases in relative antennal length, reduction in eyes (Chapter 5), and similar genetic correlation patterns among eye and antennal characters (Chapter 6).

Brandon points out that two difficulties may arise in establishing that selection has occurred. If selection has already driven the presumptive adaptative trait to fixation in the population, then active selection on this feature can no longer be observed. Indeed, this is the reason that documentation of adaptation in extremely troglomorphic cave animals is so difficult. There is little apparent morphological variation in extra-optic sensory structures, and eyes have vanished. One feature of *G. minus* is an abundance of morphological variation in troglomorphic characters, both within and among populations.

A second difficulty is that even if selection can be observed currently, it is necessary to argue that relevant parameters of the present selective environment also existed in the past. Since the caves and subsurface drainage basins predate the invasion of caves by *G. minus* (Chapter 8), so has the cave environment. The question is, how has the cave environment changed over the past million years or so? Indeed, many aspects of the cave environment have changed. As mean annual temperature rises or falls, so does the cave temperature (Chapter 4). Many of the cave streams we have studied are "underfit" from a hydrologist's point of view—the stream appears too small to have eroded the channel. Thus, at some point the cave streams were likely larger, which means there have been over time changes in resource availability. One factor has remained constant, however—the absence of light.

Ecological Explanation of Selection

While the cave environment has often been treated as a stable, unchanging one, it clearly is not (Chapter 4). Further, there is not a single "cave environment," either from the point of view of purely physical measurements or from a more general perspective. Levins and Lewontin (1985) argue effectively that the environment is defined by the organism and indeed that the distinction between organism and environment is an artificial one, an idea that is echoed in the work of some cave biologists (such as Sket 1985). However, all cave environments now and in the past share one primary feature—the total absence of light. This is the ecological theater in which selection is occurring.

The observation that directional selection occurs on eye and antennal traits of *G. minus* does not, of itself, explain the ecological advantage of large antennae and small eyes in the aphotic cave environments. For sensory structures, a plausible scenario is that a neurological tradeoff is occurring. Chemosensory and optic system inputs to the central nervous system are not processed independently, either anatomically or physiologically (Bullock and Horridge 1965). There is evidence that regions of the brain in *G. minus* (Chapter 3) are labile in terms of their relative size. It is plausible that they may also undergo changes in the source of their sensory innervation. Thus, reduction of the unused visual sensory structure (the eye) could lead to improved function of extra-optic sensory systems by decreasing the amount of "noise" in the sensory system (cf. Regal 1977) or by increasing the relative amount of the central nervous system used for processing nonvisual sensory information.

Densities of *G. minus* in caves are lower than densities in springs (Chapter 4), making mate-finding potentially more difficult. Mate-finding behavior in amphipods is under the control of pheromones, which may be sensed both by individuals in contact with and by individuals at a distance from the organism that released the pheromone (Dahl, Emanuelson, and von Mecklenberg 1970, Hartnoll and Smith 1980, Borowsky 1991). Presumptive chemoreceptors for the detection of pheromones occur on the antennae (Dahl, Emanuelson, and von Mecklenberg 1970). Lowered densities of *G. minus* in caves should lead to lowered concentrations of pheromones in the water. On the assumption that increased ability to detect sex pheromones is associated with increased antennal size, selection for increased antennal size in cave individuals would be expected.

Heritability

In a laboratory study of heritabilities under conditions of constant darkness, nearly all traits in nearly all populations showed significant heritabilities (Chapter 6 and Fong 1989), usually greater than 0.5. Thus a significant component of within-population variation is genetically determined, at least in the conditions tested. The persistence of differences between individuals from cave and spring populations reared under identical conditions implies that the among-population differences have a genetic basis as well.

Population Structure

Brandon suggests that population structure has two aspects: (1) patterns of gene flow (demic structure); and (2) patterns of selective heterogeneity (structure of the selective environment). For *G. minus* the structure of the selective environment is largely determined by the hydrologic patterns of subsurface drainage basins. The downstream resurgence populations share a selective environment of, at a minimum, light, and frequently also predation by fish. The upstream cave populations share a selective environment of, at a minimum, absence of light, and frequently also resource scarcity. The selective environments of upstream karst-window populations (Chapter 8) are in some ways intermediate, but they share with resurgences the presence of light.

Gene flow (Chapter 5) is largely occurring among cave and karst-window populations within basins and among resurgence populations. There is little if any gene flow between resurgences and upstream cave and karst-window populations. There is little if any gene flow among cave and karst window populations in different basins. This pattern of reduced gene flow has led us to suggest that what is currently called *Gammarus minus* is a complex of species, with cave populations highly convergent with one another. The evolution of troglomorphy appears to require both a separation of subsurface and surface habitats and the presence of an extensive subsurface habitat.

There is no evidence of either a genetic bottleneck in the past or current inbreeding. Ecologically and genetically, most populations are large. The only possible case of genetic bottleneck is for those populations that invaded one basin and were apparently captured by another basin as a result of subterranean stream capture (Chapter 5). All in all, the data support the view of Rouch and Danielopol (1987), who argue that for some species, cave invasions occur through active colonization rather than through passive stranding in subterranean habitats. This conclusion reinforces our view that the cave habitat is not necessarily exotic or hostile except in its absence of light.

Phylogenetic State

It is clear that troglomorphy represents the derived condition in *Gammarus minus*. Troglomorphic forms of *G. minus* are limited to two quite restricted areas embedded within its geographic range. Thus troglomorphic forms must be the result of invasion of caves by morphologically unmodified individuals from surface habitats.

The story is more complicated in the case of karst window populations. Genetic analysis suggests that they are derived from cave populations and so, for them, troglomorphy becomes the ancestral trait. The reversal in evolutionary direction in this case is equivalent to homoplasy, as is the convergence of cave populations in different basins.

Darwin's Finches and *Gammarus minus*

The finches of the Galápagos Islands are arguably the best-studied and most convincing case of natural selection and adaptive radiation known. At first glance, evolution of the Galápagos finches and evolution of *Gammarus minus* share little in common, except that the same basic processes of selection, isolation, and genetic drift influence both. There are deeper similarities than this, however, and both the similarities and the differences are informative.

The first similarity is that Darwin made little use of either cave animals or Galápagos finches in *On the Origin of Species*. While it may well be true that his reflections on finches played an important role in his formulation of the theory of natural selection, he does not explicitly mention them in his discussion of it (Grant 1986). By all accounts, Darwin was confused by the patterns of the finches, and his understanding of geographic replacement of species came not from the finches but from the mockingbirds of the Gallápagos (Sulloway 1982, Mayr 1991).

A second similarity is the age of the habitat and age of invasion. The geologic age of the Galápagos is approximately 5 million years, roughly the same age as the karst basins where troglomorphic *G. minus* occurs. While the age of the invasion of the first finch on the Galápagos is not known, largely because of disagreements about who their closest relative on the mainland is (Grant 1986, p. 253), the first speciation event (the split off of the warbler finch) is dated at 570,000 years on the basis of protein variation. This corresponds with the date of the oldest invasion of *G. minus* into caves (Chapter 8).

A third similarity is the overriding importance of directional selection in morphological change for both groups. Probably the best-documented cases of directional selection in natural populations are for bill size and shape in *Geospiza fortis* on Daphne Major Island (Grant 1986) and *Geospiza conirostris* on Isla Genovesa (Grant and Grant 1989). For both the finches and *G. minus*, the target of directional selection is not overall size of a particular character: it is shape, especially bill shape, in the case of finches, and relative antennal and eye size for *G. minus*. Furthermore, stabilizing selection appears to be less common in both groups than might be expected. For some finch species, such as *Geospiza fortis*, size is apparently stabilized by countervailing directional selection rather than stabilizing selection. The same holds for resurgence populations of *G. minus* (Chapter 6).

A fourth similarity is that at least for some physical variables, both environments are harsh. Lack (1983, p. 1) describes the Galápagos as "miles of dreary, greyish brown thornbush, in most parts dense, but sparser where there had been a more recent lava flow, and the ground still resembled a slag heap." From a more modern perspective, it is clear that Lack overestimated both the harshness of the habitat and the similarities among the islands (Ratcliffe and Boag 1983). Lack's failure to recognize the importance of floristic differences among islands led him to attribute most differences in the finches to competition, rather than adaptation to different environments.

Is it possible that we have neglected important differences among cave streams in our study of *Gammarus minus*? The diversity of subsurface habitats of which cave streams are only one has long been recognized (Racovitza 1907). Likewise, there are a variety of cave stream habitats within a cave. The most obvious examples are stream segments of different order. Cave streams range from small trickles of water with discharge of less than 0.1 l/sec to large streams with discharge of more than 10 l/sec. Poulson (1992) has probably recognized more aquatic subdivisions in cave streams than any other cave biologist. For example, he distinguishes shallow and deep base-level streams. However, all of these subdivisions are more or less repeated in every cave or at least in every basin. Differences among basins are likely to be differences in surface inputs, such as the number of surface streams within the basin (see Fig. 5.4). As far as we can determine, this factor affects the degree of isolation of the habitat more than the nature of the habitat itself. Nevertheless, subtle differences in stream habitats among basins may well exist. For example, there may be differences in channel geometry and differences in the fractal dimension of cave streams in different basins.

The degree of isolation of the habitat brings us to a major difference between Darwin's finches and *Gammarus minus*. There are thirteen species of finches, all apparently descended from a single colonizing population. The collection of populations called *Gammarus minus* may actually comprise several species (Chapter 8) but these "species" differ from the thirteen finch species in one overriding way. The "species" of *G. minus* in different basins are parallel and convergent in their morphology. While some species of finches show convergent morphology, they also show character displacement, character release, and adaptive radiation (Grant 1986).

A major, and probably the major, selective agent for Darwin's finches is interspecific competition with other finches. The scenario of the evolution of Darwin's finches involves cycles of isolation and reinvasion. What is missing in the case of *Gammarus minus* is reinvasion, and as a consequence interspecific competition is not the major selective agent. As far as we have been able to determine, there have been no cases where *G. minus* populations in one subsurface basin have invaded another subsurface basin. It is as though finches invaded each of the Galápagos islands and did not reinvade any other islands. The number of finch species on the seventeen major Galápagos islands ranges from three to ten (Grant 1986, p. 53). The number of *G. minus* "species" in each of the basins we have studied is always one.

The history of evolution and adaptation in *Gammarus minus* is clearly much simpler and more straightforward than evolution and adaptation of Darwin's finches. That of *Gammarus minus* consequently lacks the richness of pattern found in Darwin's finches. For example, there has not been adaptive radiation in *Gammarus minus*. What evolution and adaptation in *Gammarus minus* do provide, however, is a clear case of directional selection repeated in several places, resulting in convergent adaptation to the cave environment.

Selected References

Brandon, R. N. 1990. *Adaptation and environment.* Princeton, N.J.: Princeton University Press. A discussion of criteria for demonstrating that adaptation and natural selection has occurred.

Grant, P. R. 1986. *Ecology and evolution of Darwin's finches.* Princeton, N.J.: Princeton University Press. A classic study of adaptation and natural selection.

▼ Glossary

Many basic evolutionary terms refer to very complex ideas; the definitions that follow do not address the more ambiguous or contentious interpretations of the terms but instead give the more common and general meanings. We recommend *Keywords in Evolutionary Biology* (Keller and Lloyd 1992) for further information on essential evolutionary terms.

adaptation A process of genetic change resulting in improvement of a character with reference to a specific function, or a feature that has become prevalent in a population because of a selective advantage. See Chapter 9 for more on this complex concept.

adaptive radiation Evolutionary divergence of members of a single phylogenetic line into a variety of different adaptive forms.

allochthonous Originating outside the habitat under discussion.

allometry Growth of characters at different rates, resulting in changes in shape correlated with increase or decrease in size.

allopatric Used of a species or population occupying a geographic region different from that of another species or population.

allozyme One of several forms of an enzyme coded for by different alleles at a particular gene locus. See **electromorph**.

amplexus In *Gammarus,* the grasping and carrying of females by males prior to fertilization. See **precopulation**.

analogous Pertaining to traits or characters shared by different species not because of common descent but because of convergent evolution.

antennules The second pair of antennae in Crustacea.

apomorphic Pertaining to a derived character state within the group of taxa being considered, in contrast to the ancestral (plesiomorphic) state.

biplot A multivariate statistical technique that allows the simultaneous display of both observations and variables for tabular data.

bootstrap A statistical procedure involving resampling the original data in order to estimate a parameter.

bottleneck A drop in population size and, therefore, in the genetic variability of the population.

brood pouch In *Gammarus* females, four pairs of ventral plates, arising near the bases of pereopods 2–5, that become fringed with setae. Eggs are brooded within the interlocking plates.

calceoli In some amphipods, male antennular organs that are not innervated and have a plate-like structure.

caprock In karst areas, an impermeable rock, typically sandstone, that covers limestone.

cladistic Pertaining to branching patterns; a cladistic classification classifies organisms on the basis of the historical sequences by which they have diverged from common ancestors.

commensalism An ecological relationship between species in which one is benefited but the other is little affected.

conduit flow Subsurface water movement in large openings (cave passages), analogous to irregular pipes.

convergence The independent evolution of similar traits in two or more genetically distinct taxa by different genetic and developmental pathways.

deme A local population.

detritivores Organisms that feed on decaying organic material.

diffuse flow Subterranean water movement in porous media such as sand; in contrast to **conduit flow**.

directional selection Selection for a higher (or lower) value of a character than its current mean value.

dispersalism Speciation as a result of dispersal; also termed active allopatric speciation.

disruptive selection Selection for two modal phenotypes and against those intermediate between them.

divergence The evolution of dissimilar traits in different species sharing a common ancestor species.

Dollo's law Complex structures, once lost, are not regained in their original form.

downcutting An eroding stream.

ecotone The boundary between two communities or habitats.

electromorph An electrophoretically distinguishable form of a protein, assumed to represent a unique allele or group of alleles.

epigean The surface environment as opposed to the subsurface or subterranean environment.

epigenesis Development, especially pertaining to interactions among developmental processes above the level of primary gene action.

evolution In biological evolution, the change in gene frequencies over time.

evolutionary history The phylogenetic history of a taxa or a set of taxa.

evolutionary tradeoff The elaboration or acquisition of a trait or characteristic made possible by a loss or diminution of another trait or characteristic, both changes having a genetic basis.

exaptation Adaptation for one function serving for another function.

fractal dimension The similarity dimension of geometric figures that are self-similar at different scales. In general, fractal dimensions are used to quantify a subjective feeling about how densely the object occupies the space in which it lies.

F-statistics A comparative measure of intraspecific genetic differentiation.

gene flow The incorporation of genes from one population into another population.

genetic drift Random changes in gene frequency within a population.

genotype The genes possessed by an organism.

heritability The proportion of the variance among individuals for a trait that is attributable to differences in genotype.

Holarctic zone North Temperate zone.

homologous Pertaining to traits that have evolved from a common ancestral trait.

homoplasy Possession by two or more species of a similar trait that has not been derived by both species from their common ancestor; causes may be convergence, parallelism, or evolutionary reversal.

karst A landform type characterized by sinkholes, caves, and springs developed in calcareous rock by processes of solution, not erosion.

karst window A section of subterranean stream exposed by roof collapse; more generally, short reaches of stream on the surface in karstlands.

Lamarckism Conventionally refers to the evolution of acquired characters although Lamarck himself attached greater importance to **orthogenesis**.

laminar flow Movement of water molecules in threads parallel to the direction of flow; in contrast to turbulent flow.

lotic Pertaining to running-water environments.

monotonic A function having the property either of never increasing or never decreasing as the independent variable increases.

N Population size.

N_e Effective population size, used in population genetics to approximate the size of a randomly interbreeding population that would result in the same level of inbreeding as that in the observed population.

natural selection A process that occurs if and only if the population has phenotypic variation, fitness variation, and a genetic basis for phenotypic variation. See Chapters 6 and 9 for more information.

neo-Darwinism Dating from the 1930s, the reconciliation of Darwin's theory of evolution with the facts of genetics.

neo-Davisian Modern version of the theory by geologist William Morris Davis of erosional cycles of the landscape.

neoteny The retention of juvenile characters in a sexually mature organism.

neutral mutation A genetic mutation that gives no advantage or disadvantage to the organism.

ommatidium The individual unit of a compound eye; plural, *ommatidia*.

ontogeny Growth and development of an organism from embryo to adult.

orthogenesis Evolution toward a "perfect form," determined by factors internal to the organism (see Chapter 2).

parallelism The independent evolution of similar traits in two or more genetically connected taxa by similar genetic and developmental pathways.

parapatric Used of species or populations that have contiguous but non-overlapping geographic distributions.

path analysis A multivariate statistical technique that is based on a diagram of potential causal pathways between variables.

PCA Principal-component analysis, a multivariate statistical technique that extracts independent linear combinations of variables.

pereopod Thoracic appendage of crustaceans. In amphipods, there are seven pairs of thoracic appendages, the first two of which may be termed gnathopods.

phenotype The structural, physiological, biochemical, behavioral, and other characteristics of an organism.

phreatic loop An inclined, subterranean tube formed by water moving up the tube under pressure.

phylogeny The genealogy of a group of taxa such as species.

pleiotropic Pertaining to a gene that has more than one phenotypic effect.

polje Large, flat-floored karst depression that floods periodically, with springs on one side and swallets on the other.

polytomy A node in a phylogenetic tree that has more than two descendant nodes.

polytypic Used of species with distinct geographic races or subspecies.

pre-adaptation Possession by an organism of the necessary properties to permit a shift into a new niche or habitat. A structure is preadapted if it can assume a new function before it becomes modified itself.

precopula In *Gammarus*, the carrying of females by males (see **amplexus**) prior to fertilization.

RAPD Random amplification of polymorphic DNA; DNA amplified via polymerase chain reaction using ten base primers of random sequence.

refugium A refuge either from climatic vicissitudes or from predators and competitors.

regressive evolution The loss of morphological and behavioral characters that accompanies isolation in caves.

resurgence The re-emergence at the surface of a watercourse that originated as a surface stream on impermeable rock and then passed underground through caves and fissures.

selection Nonrandom differential survival or reproduction of classes of entities in successive generations; see **natural selection**.

speciation Formation of new species.

species The basal taxonomic unit in biological classification. Usually defined on the basis of common evolutionary fate (the phylogenetic species concept) or on the basis of the potential for interbreeding under natural conditions (the biological species concept). See Chapter 7.

speciose Used of genera or families containing many species.

splines A non-parametric regression technique that allows an objective evaluation of the shape of a function.

stabilizing selection Selection against phenotypes that deviate in either direction from an optimal value.

stream order A technique of stream classification based on patterns of tributaries. The commonest classification (Horton) assigns tributaries with no branches first-order; streams receiving only first-order streams are second-order; streams receiving only second- and lower-order streams are third-order; and so on.

stream piracy The capture of one stream by another as a result of differential erosion of stream channels.

swallet The point where a stream disappears underground.

sympatric Used of two or more species or populations occupying the same goegraphic locality.

synapomorphy A derived (apomorphic) state shared by two or more taxa.

troglobite Species that completes its entire life cycle inside caves and is never found in surface habitats.

troglomorphic Pertaining to morphological and behavioral characters that are convergent in subterranean populations.

troglophile Species that can complete its entire life cycle inside caves but is also found outside of caves.

trogloxene Species that cannot complete its entire life cycle in caves.

underfit Used of a stream that appears too small to have eroded the channel in which it flows.

UPGMA Unweighted pair-group method using arithmetic averages, a technique for clustering data and producing dendrograms.

uropod In *Gammarus,* an appendage of the last three abdominal segments.

urosome In *Gammarus,* the last three body segments of the abdominal region.

vicariance Speciation as a result of range disruption, typically the result of some non-biological process; also termed passive allopatric speciation.

▼ References

Anders, F. 1956. Über Ausbildung und Vererbung der Körperfarbe bei *Gammarus pulex* ssp. *subterraneus* (Schneider), einer normalerweise pigmentlosen Höhlenform des gemeinen Bachflohkrebses. *Zeitschrift für induktiv Abstammungs- und Vererbungslehre* 87:567–579.

Arnold, S. J. 1983. Morphology, performance and fitness. *American Zoologist* 23:347–361.

Banta, A. M. 1907. The fauna of Mayfield's Cave. *Carnegie Institute of Washington Publications* 67:1–114.

Bärlocher, F., and B. Kendrick. 1973. Fungi in the diet of *Gammarus pseudoliminaeus* (Amphipoda). *Oikos* 24:295–300.

—— 1975. Assimilation efficiency of *Gammarus pseudoliminaeus* (Amphipoda) feeding on fungal mycelium or autumn-shed leaves. *Oikos* 26:55–59.

Barnard, J. L., and C. M. Barnard. 1983. *Freshwater Amphipoda of the world.* 2 vols. Mount Vernon, Va.: Hayfield Assoc.

Barr, T. C. 1966. Evolution of cave biology in the United States, 1822–1965. *National Speleological Society Bulletin* 28:15–21.

—— 1968. Cave ecology and the evolution of troglobites. *Evolutionary Biology* 2:35–102.

—— 1979. The taxonomy, distribution, and affinities of *Neaphaenops*, with notes on associated species of *Pseudanophthalmus* (Coleoptera, Carabidae). *American Museum Novitates* no. 2682.

—— 1985. Pattern and process in speciation of trechine beetles in eastern North America (Coleoptera: Carabidae: Trechinae). In *Taxonomy, phylogeny, and zoogeography of beetles and ants,* ed. G. E. Ball, pp. 350–407. Leiden: W. Junk.

Barr, T. C., and J. R. Holsinger. 1985. Speciation in cave faunas. *Annual Review of Ecology and Systematics* 16:313–337.

Barr, T. C., and R. A. Kuehne. 1971. Ecological studies in the Mammoth Cave system of Kentucky. II. The ecosystem. *Annales de Speleologie* 26:47–96.

Barton, N. H. 1990. Pleiotropic models of quantitative variation. *Genetics* 124:773–782.

Beatty, R. A. 1949. The pigmentation of cavernicolous animals. III. The carotenoid pigments of some amphipod Crustacea. *Journal of Experimental Biology* 26:125–130.

Bellés, X. 1992. From dragons to allozymes. A brief account of the history of biospeleology. In *The natural history of biospeleology*, ed. A. Camacho, pp. 3–24. Madrid: Museo Nacional de Ciencias Naturales.

Bookstein, F. L. 1989. "Size and shape": a comment on semantics. *Systematic Zoology* 38:173–180.

Borowsky, B. 1984. The use of the male's gnathopods during precopulation in some gammaridean amphipods. *Crustaceana* 47:245–250.

—— 1991. Patterns of reproduction of some amphipod crustaceans and insights into the nature of their stimuli. In *Crustacean sexual biology*, ed. R. T. Bauer and J. W. Martin, pp. 33–49. New York: Columbia University Press.

Botosaneanu, L., ed. 1986. *Stygofauna mundi*. Leiden: E. J. Brill.

Bousfield, E. L. 1958. Fresh-water amphipod crustaceans of glaciated North America. *Canadian Field Naturalist* 72:55–113.

Boutin, C., M. Messouli, and N. Coineau. 1992. Phylogénie et biogéographie évolutive d'un groupe de Metacrangonyctidae, Crustacés Amphipodes stygobies du Maroc. II—Cladistique et paléobiogéographie. Avec l'examen comparatif de plusieurs logiciels de parcimonie. *Stygologia* 7:159–178.

Bowler, P. J. 1983. *The eclipse of Darwinism*. Baltimore: Johns Hopkins University Press.

Brandon, R. N. 1990. *Adaptation and environment*. Princeton, N.J.: Princeton University Press.

Bruce, R. C. 1979. Evolution of paedomorphosis in salamanders in the genus *Gyrinophilus*. *Evolution* 33:998–1000.

Bullock, T. H., and G. A. Horridge. 1965. *Structure and function in the nervous system of invertebrates*, vol. 2. San Francisco: Freeman.

Caccone, A., and J. Powell. 1987. Molecular evolutionary divergence among North American cave crickets. II. DNA–DNA hybridization. *Evolution* 41:1215–1238.

Cavalli-Sforza, L. L., and A. W. F. Edwards. 1967. Phylogenetic analysis: models and estimation procedures. *Evolution* 21:550–570.

Cain, A. J., and P. M. Sheppard. 1954. Natural selection in *Cepaea*. *Genetics* 39:89–116.

Chakraborty, R., and M. Nei. 1974. Dynamics of gene differentiation between incompletely isolated populations of unequal sizes. *Theoretical Population Biology* 5:460–469.

Cheverud, J. M. 1982. Phenotypic, genetic, and environmental integration in the cranium. *Evolution* 36:499–516.

Christiansen, K. A. 1961. Convergence and parallelism in cave Entomobryinae. *Evolution* 15:288–301.
——— 1962. Proposition pour la classification des animaux cavernicoles. *Spelunca* 2:76–78.
——— 1965. Behaviour and form in the evolution of cave Collembola. *Evolution* 19:529–532.
——— 1992. Cave life in the light of modern evolutionary theory. In *The natural history of biospeleology*, ed. A. Camacho, pp. 453–478. Madrid: Museo Nacional de Ciencias Naturales.
Christiansen, K. A., and D. C. Culver. 1968. Geographical variation and evolution in *Pseudosinella hirsuta*. *Evolution* 22:237–255.
Cockerum, C. C. 1969. Variance of gene frequencies. *Evolution* 23:72–84.
Cole, G. A. 1980. The mandibular palps of North American freshwater species of *Gammarus* (Crustacea: Amphipoda). *Crustaceana Supplement* 6:68–83.
Cooper, J. E. 1975. *Ecological and behavioral studies in Shelta Cave, Alabama, with emphasis on decapod crustaceans*. Ph.D. diss., University of Kentucky, Lexington, Ky.
Cope, E. D. 1868. On the origin of genera. *Proceedings of the Academy of Natural Sciences of Phildelphia* 20:242–300.
——— 1872. On the Wyandotte Cave and its fauna. *American Naturalist* 6:406–422.
Crouau-Roy, B. 1988. Genetic structure of cave-dwelling beetle populations: significant deficiencies of heterozygotes. *Heredity* 60:321–327.
Culver, D. C. 1971. Analysis of simple cave communities. III. Control of abundance. *American Midland Naturalist* 85:173–187.
——— 1975. The interaction of competition and predation in cave stream communities. *International Journal of Speleology* 7:229–245.
——— 1976. The evolution of aquatic cave communities. *American Naturalist* 110:949–957.
——— 1982. *Cave life: Evolution and ecology*. Cambridge, Mass.: Harvard University Press.
——— 1985. Trophic relationships in aquatic cave environments. *Stygologia* 1:43–53.
——— ed. 1985. *Regressive evolution*. National Speleological Society Bulletin 47, no. 2.
——— 1987. Eye morphometrics of cave and spring populations of *Gammarus minus* (Amphipoda: Gammaridae). *Journal of Crustacean Biology* 7:136–147.
Culver, D. C., D. W. Fong, and R. W. Jernigan. 1991. Species interactions in cave stream communities: experimental results and microdistribution effects. *American Midland Naturalist* 126:364–379.

Culver, D. C., and J. R. Holsinger. 1992. How many species of troglobites are there? *National Speleological Society Bulletin* 54:79–80.

Culver, D. C., and T. L. Poulson. 1971. Oxygen consumption and activity in closely related amphipod populations from cave and surface habitats. *American Midland Naturalist* 85:74–84.

Culver, D. C., R. W. Jernigan, J. O'Connell, and T. C. Kane. 1994. The geometry of natural selection in cave and spring populations of the amphipod *Gammarus minus* Say (Crustacea: Amphipoda). *Biological Journal of the Linnean Society* 52:49–67.

Culver, D. C., W. K. Jones, D. W. Fong, and T. C. Kane. In press. Organ Cave karst basin. In *Groundwater ecology*, ed. J. Gibert and J. Stanford. New York: Academic Press.

Curl, R. 1966. Caves as a measure of karst. *Journal of Geology* 74:798–830.

—— 1988. Fractal dimensions and geometries of caves. *Mathematical Geology* 18:765–783.

Dahl, E., H. Emanuelson, and C. von Mecklenberg. 1970. Pheromone transport and reception in an amphipod. *Science* 170:739–740.

Dai, N. 1989. *The comparative examination of eye anatomy of spring-dwelling and cave-dwelling populations of the amphipod* Gamarus minus. M.S. thesis, American University, Washington, D.C.

Danielopol, D. L., and R. Rouch. 1991. L'adaptation des organismes au milieu aquatique souterrain. Réflexions sur l'apport des recherches écologiques récentes. *Stygologia* 6:129–142.

Darwin, C. 1859. *On the origin of species by means of natural selection, or the preservation of favoured races in the struggle for life.* [First ed.] London: John Murray.

—— 1872. *On the origin of species by means of natural selection, or the preservation of favoured races in the struggle for life.* 6th ed. London: John Murray.

Dasher, G., and W. Balfour. 1994. *The caves of the Buckeye Creek drainage, Greenbrier County, West Virginia.* Bulletin of the West Virginia Speleological Survey, Barrackville, W.Va.

Deike, G. H. 1988. Geological factors in the hydrology of the Organ Cave plateau, West Virginia, USA. *Proceedings 21st Congress, International Association of Hydrogeologists, Guilin, China* 1:394–399.

Delong, M. D. 1992. A new species of *Gammarus* (Crustacea: Amphipoda: Gammaridae) from the Lower Mississippi River, Louisiana. *American Midland Naturalist* 127:241–247.

DeSalle, R., T. Freedman, E. M. Prager, and A. C. Wilson. 1987. Tempo and mode of sequence evolution in mitochondrial DNA of Hawaiian *Drosophila*. *Journal of Molecular Evolution* 26:157–164.

Douglas, M. E., and J. A. Endler. 1982. Quantitative matrix comparisons in ecological and evolutionary investigations. *Journal of Theoretical Biology* 99:777–795.

Dunham, P. J., T. Alexander, and A. Hurshman. 1986. Precopulatory mate guarding in an amphipod, *Gammarus lawrencianus* Bousfield. *Animal Behaviour* 34:1680–1686.

Elwood, R. W., and J. T. A. Dick. 1990. The amorous *Gammarus:* the relationship between precopula duration and size assortative mating in *G. pulex. Animal Behaviour* 39:828–833.

Elwood, J. W., J. D. Newbold, R. V. O'Neill, and W. Van Winkle. 1983. Resource spiralling: an operational paradigm for analyzing lotic ecosystems. In *Dynamics of lotic ecosystems,* ed. T. D. Fontaine and S. M. Bartell, pp. 3–27. Ann Arbor, Mich.: Ann Arbor Science Publishers.

Elwood, J. W., J. D. Newbold, R. V. O'Neill, R. W. Stark, and P. T. Singley. 1981. The role of microbes associated with organic and inorganic substrates in phosphorus spiralling in a woodland stream. *Verhandlungen Internationale Vereinigung der Limnologie* 21:818–824.

Endler, J. A. 1986. *Natural selection in the wild.* Princeton, N.J.: Princeton University Press.

Falconer, D. W. 1989. *Introduction to quantitative genetics.* Essex, England: Longman.

Felsenstein, J. 1985. Confidence limits on phylogenies: an approach using bootstrap. *Evolution* 39:783–791.

Fisher, R. A. 1930. *The genetical theory of natural selection.* London: Blackwell.

Fong, D. W. 1985. *A quantitative genetic analysis of regressive evolution in the amphipod* Gammarus minus *Say.* Ph.D. diss., Northwestern University, Evanston, Ill.

——— 1989. Morphological evolution of the amphipod *Gammarus minus* in caves: quantitative genetic analysis. *American Midland Naturalist* 121:361–378.

Fong, D. W., and D. C. Culver. 1985. A reconsideration of Ludwig's differential migration theory of regressive evolution. *National Speleological Society Bulletin* 47:123–127.

Fong, D. W., and D. C. Culver. 1994. Fine-scale biogeography of the crustacean fauna of Organ Cave, West Virginia. *Hydrobiologia* 287:29–37.

Ford, D. C., and R. O. Ewers. 1978. The development of limestone cave systems in the dimensions of length and breadth. *Canadian Journal of Earth Science* 15:1783–1798.

Ford, D. C., and P. W. Williams. 1989. *Karst geomorphology and hydrology.* London: Unwin Hyman.

Gabriel, K. R. 1971. The biplot-graphic display of matrices with application to principal component analysis. *Biometrika* 58:453–467.

Gibert, J., M-J. Dole-Olivier, P. Marmonier, and P. Vervier. 1990. Surface water–groundwater ecotones. In *The ecology and management of aquatic-terrestrial ecotones,* ed. R. J. Naiman and H. Decamps, pp. 193–225. Princeton, N.J.: Parthenon.

Gillespie, J. H. 1991. *The causes of molecular evolution.* New York: Oxford University Press.

Glazier, D. S., M. T. Horne, and M. E. Lehman. 1992. Abundance, body composition and reproductive output of *Gammarus minus* (Crustacea: Amphipoda) in ten cold springs differing in pH and ionic content. *Freshwater Biology* 28:149–163.

Godfrey, R. B., J. R. Holsinger, and K. A. Carson. 1988. A comparison of the morphology of calceoli in the freshwater amhipods *Crangonyx richmondensis* s. lat. (Crangonyctidae) and *Gammarus minus* (Gammaridae). *Crustaceana* 13:115–121.

Goedmakers, A. 1980. Microgeographic races of *Gammarus fossarum* Koch, 1836. *Crustaceana Supplement* 6:216–224.

Gooch, J. L., and D. S. Glazier. 1991. Temporal and spatial patterns in mid-Appalachian springs. *Memoirs of the Entomological Society of Canada* 155:29–49.

Gould, S. J. 1977. *Ontogeny and phylogeny.* Cambridge, Mass.: Harvard University Press.

Grant, B. R., and P. R. Grant. 1989. Natural selection in a population of Darwin's finches. *American Naturalist* 133:377–393.

Grant, P. R. 1986. *Ecology and evolution of Darwin's finches.* Princeton, N.J.: Princeton University Press.

Gulden, R. 1993. U.S.A. long cave list. In *1993 Members Manual,* ed. S. Fee and J. Fee, p. 190. Huntsville, Ala.: National Speleological Society.

Hack, J. T. 1960. Interpretation of erosional topography in humid temperate regions. *American Journal of Science* 258A:80–97.

Haeckel, E. 1874. *Anthropogenie: Keimes- und Stammes-Geschichte des Menschen.* Leipzig: W. Engelmann.

Hamilton-Smith, E. 1971. The classification of cavernicoles. *National Speleological Society Bulletin* 33:63–66.

Hargrave, B. T. 1970. The utilization of benthic microflora by *Hyallela azteca* (Amphipoda). *Journal of Animal Ecology* 39:427–437.

Hartnoll, R. J., and S. M. Smith. 1978. Pair formation and the reproductive cycle in *Gammarus duebenii. Journal of Natural History* 12:501–511.

Heller, S. A. 1991. An attempt to model an Appalachian karst aquifer using MODFLOW. In *Appalachian karst,* ed. E. H. Kastning and K. M. Kastning, pp. 177–186. Huntsville, Ala.: National Speleological Society.

Henry, J. P., and G. Magniez. 1983. Introduction pratique a la systematique des organismes des eaux continentales Françaises. 4. Crustacés Isopodes (principalement Asellotes). *Bulletin mensuel de la Société Linnéenne* de Lyon 52:319–358.

Hershler, R., J. R. Holsinger, and L. Hubricht. 1990. A revision of the North American snail genus *Fontigens* (Prosobranchia: Hydrbiidae). *Smithsonian Contributions to Zoology*, no. 509.

Hillis, D. M., and C. Moritz, eds. 1990. *Molecular systematics.* Sunderland, Mass.: Sinauer.

Holomuzki, J. R., and J. D. Hoyle. 1990. Effect of predatory fish presence on habitat use and diel movement of the stream amphipod, *Gammarus minus. Freshwater Biology* 24:509–517.

Holsinger, J. R. 1972. *The freshwater amphipod crustaceans (Gammaridae) of North America.* Biota of Freshwater Ecosystems Identification Manual No. 5. Washington, D.C.: Environmental Protection Agency.

——— 1975. *Descriptions of Virginia caves.* Bulletin 85. Charlottesville, Va.: Virginia Division of Mineral Resources.

——— 1986. Amphipoda: Holarctic crangonyctid amphipods. In *Stygofauna mundi,* ed. L. Botosaneanu, pp. 535–549. Leiden: E. J. Brill.

——— 1988. Troglobites: the evolution of cave-dwelling organisms. *American Scientist* 88:146–155.

Holsinger, J. R., R. A. Baroody, and D. C. Culver. 1976. *The invertebrate cave fauna of West Virginia.* Bulletin 5. Barrackville, W.Va.: West Virginia Speleological Survey.

Holsinger, J. R., and D. C. Culver. 1970. Morphological variation in *Gammarus minus* Say (Amphipoda, Gammaridae), with emphasis on subterranean forms. *Postilla,* no. 146.

Houle, D. 1991. Genetic covariance of fitness correlates: what genetic correlations are made of and why it matters. *Evolution* 45:630–648.

Hovey, H. C. 1882. *Celebrated American caverns.* Cincinnati: R. Clarke and Co.

Howarth, F. G. 1980. The zoogeography of specialized cave animals: a bioclimatic model. *Evolution* 34:394–406.

——— 1993. High-stress subterranean habitats and evolutionary change in cave-inhabiting arthropods. *American Naturalist* 142:S65–S77.

Hubricht, L. 1943. Studies in the Nearctic freshwater Amphipoda, III. Notes on the freshwater Amphipoda of eastern United States, with descriptions of ten new species. *American Midland Naturalist* 29:683–712.

Hyatt, A. 1866. On the parallelism between different stages of life in the individual and those in the entire group of the Molluscous order Tetrabranchiata. *Memoirs of the Boston Society of Natural History* 1:193–209.

Hynes, H. B. N. 1955. The reproductive cycle of some British freshwater Gammaridae. *Journal of Animal Ecology* 24:352–387.

Janzer, W., and W. Ludwig. 1952. Versuche zur evolutorischen Entstehung der Hohlentiermerkmale. *Zeitschrift für inductive Abstammungs- und Vererbungslehre* 84:462–479.

Jeannel, R. 1923. Sur l'évolution des Coléoptères aveugles et le peuplement des grottes dan les monts du Bihor, en Transylvanie. *Compte Rendus Academie Sciences Paris* 176:1670–1673.

Jernigan, R. W., D. C. Culver, and D. W. Fong. 1994. The dual role of selection and evolutionary history as reflected in genetic correlations. *Evolution*, vol. 48 (forthcoming).

Jones, R. 1990. *Evolution of cave and surface populations of the amphipod* Gammarus minus. Ph.D. diss., Northwestern University, Evanston, Ill.

Jones, R., and D. C. Culver. 1989. Evidence for selection on sensory structures in a cave population of *Gammarus minus* Say (Amphipoda). *Evolution* 43:688–693.

Jones, R., D. C. Culver, and T. C. Kane. 1992. Are parallel morphologies of cave organisms the result of similar selection pressures? *Evolution* 46:353–365.

Jones, W. K. 1973. *Hydrology of limestone karst in Greenbrier County, West Virginia.* Bulletin 36. Morgantown, W.Va.: West Virginia Geologic and Economic Survey.

Juberthie, C. 1989. Insularité et speciation souterraine. *Mémoires de Biospéologie* 16:3–14.

Kane, T. C., T. C. Barr, and W. J. Badaracca. 1992. Cave beetle genetics: geology and gene flow. *Heredity* 68:277–286.

Kane, T. C., and G. D. Brunner. 1986. Geographic variation in the cave beetle *Neaphaenops tellkampfi* (Coleoptera: Carabidae). *Psyche* 93:231–251.

Kane, T. C., D. C. Culver, and R. T. Jones. 1992. Genetic structure of morphologically differentiated populations of the amphipod *Gammarus minus*. *Evolution* 46:272–278.

Kane, T. C., and R. C. Richardson. 1985. Regressive evolution: an historical perspective. *National Speleological Society Bulletin* 47:71–77.

Karaman, G. S., and S. Pinkster. 1977. Freshwater *Gammarus* species from Europe, North Africa, and adjacent regions of Asia (Crustacea—Amphipoda) Part I. *Gammarus pulex* group and related species. *Bijdragen tot der Dierkunde* 47:1–97.

Karaman, G. S., and S. Ruffo. 1986. Amphipoda: *Niphargus*-group (Niphargidae sensu Bousfield 1982). In *Stygofauna mundi*, ed. L. Botosaneanu, pp. 514–534. Leiden: E. J. Brill.

Kaushik, N. K., and H. B. N. Hynes. 1971. The fate of dead leaves that fall into streams. *Archiv für Hydrobiologie* 68:465–515.

Kimura, M. 1983. *The neutral theory of molecular evolution.* New York: Cambridge University Press.

Kosswig, C. 1965. Génétique et évolution régressive. *Revue des Questions Scientifiques* 136:227–257.

Kosswig, C., and L. Kosswig. 1940. Die Variabilität bei *Asellus aquaticus* unter besonderer Berucksichtigung der Variabilität in isolierten unter- und oberirdischen Populationen. *Revue de Facultie des Sciences* (Istanbul), ser. B, 5:1–55.

Kostalos, M. S. 1979. Life history and ecology of *Gammarus minus* Say (Amphipoda, Gammaridae). *Crustaceana* 37:113–122.

Kostalos, M. S., and R. L. Seymour. 1976. Role of microbial enriched detritus in the nutrition of *Gammarus minus* (Amphipoda). *Oikos* 27:512–516.

Lack, D. L. 1983. *Darwin's finches*. 2d ed. Cambridge: Cambridge University Press.

Laing, C. D., G. R. Carmody, and S. B. Peck. 1976a. How common are sibling species in cave-inhabiting invertebrates? *American Naturalist* 110:184–189.

——— 1976b. Population genetics and evolutionary biology of the cave beetle *Ptomaphagus hirtus*. *Evolution* 30:484–494.

Lamarck, J. B. 1984. *Zoological philosophy: An exposition with regard to the natural history of animals*. Trans. H. Elliot. Chicago: University of Chicago Press.

LaMotte, M. 1951. Recherches sur la structure genetique des populations naturelles de *Cepaea nemoralis* (L.). *Bulletin Biologique de France et Belgique, supplement* 35:1–238.

Lande, R. 1976. Natural selection and random genetic drift in phenotypic evolution. *Evolution* 30:314–334.

——— 1979. Quantitative genetic analysis of multivariate evolution, applied to brain-body size allometry. *Evolution* 33:402–416.

——— 1982. A quantitative genetic theory of life history evolution. *Ecology* 63:607–615.

Lande, R., and S. J. Arnold. 1983. The measurement of selection on correlated characters. *Evolution* 37:1210–1226.

Langecker, T. G., H. Schmale, and H. Wilkens. 1993. Transcription of the opsin gene in degenerate eyes of cave-dwelling *Astyanax fasciatus* (Teleostei, Characidae) and of its conspecific epigean ancestor during early ontogeny. *Cell and Tissue Research* 273:183–192.

Levins, R., and R. C. Lewontin. 1985. *The dialectical biologist*. Cambridge, Mass.: Harvard University Press.

Lewontin, R. C. 1972. *The genetic basis of evolutionary change*. New York: Columbia University Press.

——— 1974. The analysis of variance and the analysis of causes. *American Journal of Human Genetics* 26:400–411.

Ludwig, W. 1942. Zur evolutorischen Erklärung der Höhlentiermerkmale durch Allelemination. *Biologisches Zentralblatt* 62:447–482.

Lynch, M. 1988. The rate of polygenic mutation. *Genetical Research* 51:137–148.

——— 1990. The rate of morphological evolution in mammals from the standpoint of neutral expectation. *American Naturalist* 136: 727–741.

——— 1991. Methods for the analysis of comparative data in evolutionary biology. *Evolution* 45:1065–1080.

Macfadyen, A. 1961. Metabolism of soil invertebrates in relation to soil fertility. *Annals of Applied Biology* 49:215–218.

MacQueen, J. 1967. Some methods for classification and analysis of multivariate observations. In *Proceedings of the Fifth Berkeley Symposium on Mathematical Statistics and Probability*, pp. 281–297. Berkeley: University of California Press.

Magniez, G. 1985. Regressive evolution in stenasellids (Crustacea, Isopoda, Asellota) of underground waters. *National Speleological Society Bulletin* 47:118–122.

Man, Z. 1991. *Life history variation among spring-dwelling populations of* Gammarus minus *Say*. M.S. thesis, American University, Washington, D.C.

Manly, B. F. J. 1991. *Randomization and Monte Carlo methods in biology*. London: Chapman and Hall.

Marchant, R. 1981. The ecology of *Gammarus* in running water. In *Perspectives in running water ecology*, ed. M. A. Lock and D. D. Williams, pp. 225–249. New York: Plenum Press.

Marchant, R., and H. B. N. Hynes. 1981. Field estimates of feeding rate for *Gammarus pseudolimnaeus* (Crustacea: Amphipoda) in the Credit River, Ontario. *Freshwater Biology* 11:27–36.

Mayr, E. 1991. *One long argument*. Cambridge, Mass.: Harvard University Press.

Milstead, B., and S. T. Threlkeld. 1986. An experimental analysis of darter predation on *Hyalella azteca* using semipermeable enclosures. *Journal of the North American Benthological Society* 5:311–318.

Minckley, W. L., and G. A. Cole. 1963. Ecological and morphological studies on gammarid amphipods (*Gammarus* spp.) in spring-fed streams of northern Kentucky. *Occasional Papers of the C. C. Adams Center for Ecological Studies* 10:1–35.

Mitchell, R. W., W. H. Russell, and W. R. Elliott. 1977. Mexican eyeless characin fishes, genus *Astyanax:* environment, distribution and evolution. Special Publication No. 12. Lubbock: Texas Tech University Museum.

Mitchell-Olds, T. 1989. *FRE-STAT user's manual*. Technical Bulletin No. 101. Missoula, Mont.: Division of Biological Sciences, University of Montana.

Mitchell-Olds, T., and R. G. Shaw. 1987. Regression analysis of natural selec-

tion: statistical inference and biological interpretation. *Evolution* 41:1149–1161.
Moore, J. W. 1975. The role of algae in the diet of *Asellus aquaticus* L. and *Gammarus pulex* L. *Journal of Animal Ecology* 44:719–730.
Motas, C. 1958. Freatobiologia, o noua ramura a limnologei. *Natura* (Bucharest) 10:95–105.
Mylroie, J. E., and J. L. Carew. 1986. Minimum duration for speleogenesis. *Proceedings Ninth International Congress of Speleology, Barcelona* 1:249–251.
Naylor, C., and J. Adams. 1987. Sexual dimorphism, drag constraints and male performance in *Gammarus duebeni* (Amphipoda). *Oikos* 48:23–27.
Nei, M. 1987. *Molecular evolutionary genetics*. New York: Columbia University Press.
Nevo, E., A. Beiles, and R. Ben-Shlomo. 1984. The evolutionary significance of genetic diversity: ecological, demographic and life history correlates. In *Evolutionary dynamics of genetic diversity*, ed. G. S. Mani, pp. 13–213. Berlin: Springer-Verlag.
Newman, R. M., and T. F. Waters. 1984. Size-selective predation on *Gammarus pseudolimnaeus* by trout and sculpins. *Ecology* 65:1535–1545.
Notenboom, J. 1988. Phylogenetic relationships and biogeography of the groundwater-dwelling amphipod genus *Pseudoniphargus* (Crustacea), with emphasis on Iberian species. *Bijdragen tot der Dierkunde* 58:159–204.
Nychka, D. 1991. Choosing a range for the amount of smoothing in nonparametric regression. *Journal of the American Statistical Association* 86:653–664.
Packard, A. S. 1888. The cave fauna of North America, with remarks on the anatomy of brain and the origin of the blind species. *Memoirs of the National Academy of Sciences (USA)* 4:1–156.
Palmer, A. N. 1989. Geomorphic history of the Mammoth Cave system. In *Karst hydrology: Concepts from the Mammoth Cave area*, ed. W. B. White and E. L. White, pp. 317–337. New York: Van Nostrand Reinhold.
Parzefall, J. 1992. Behavioural aspects in animals living in caves. In *The natural history of biospeleology*, ed. A. Camacho, pp. 327–376. Madrid: Museo Nacional de Ciencias Naturales.
Payne, F. 1911. *Drosophila ampelophila* bred in the dark for sixty-nine generations. *Biological Bulletin* 21:297–301.
Peck, S. B. 1980. Climatic change and the evolution of cave invertebrates in the Grand Canyon, Arizona. *National Speleological Society Bulletin* 42:53–60.
Petre-Stroobants, G. 1982. Analyse comparative de la variabilité de certains caractères taxonomiques de *Gammarus pulex* (Linnaeus, 1758), *Gammarus fossarum* Koch, 1835 et *Gammarus caparti* Petre-Stroobants,

1980 (Crustacea—Amphipoda). *Polski Archiwum Hydrobiologii* 29:205–219.

Pinkster, S. 1983. The value of morphological characters in the taxonomy of *Gammarus*. *Beaufortia* 33:15–28.

Poulson, T. L. 1963. Cave adaptation in amblyopsid fishes. *American Midland Naturalist* 70:257–290.

——— 1992. The Mammoth cave ecosystem. In *The natural history of biospeleology,* ed. A. Camacho, pp. 569–611. Madrid: Museo Nacional de Ciencias Naturales.

Poulson, T. L., and W. B. White. 1969. The cave environment. *Science* 165:971–981.

Prout, T. 1964. Observations on structural reduction in evolution. *American Naturalist* 98:239–249.

Racovitza, E. G. 1907. Essai sur les problèmes biospéologiques. *Archives de Zoologie Expérimentale et Générale* 4:371–488.

Ratcliffe, L. M., and P. T. Boag. 1983. Foreword to *Darwin's Finches,* by David Lack, 2nd ed. Cambridge: Cambridge University Press.

Read, A. T., and D. D. Williams. 1990. The role of calceoli in precopulatory behaviour and mate recognition of *Gammarus pseudolimnaeus* Bousfield (Crustacea, Amphipoda). *Journal of Natural History* 24:351–359.

Regal, P. 1977. Evolutionary loss of useless features: is it noise suppression? *American Naturalist* 111:123–133.

Richardson, R. C., and T. C. Kane. 1988. Orthogenesis and evolution in the 19th century: the idea of progress in American neo-Lamarckism. In *Evolutionary progress,* ed. M. Nitecki, pp. 149–167. Chicago: University of Chicago Press.

Rohlf, F. J., and F. J. Bookstein, eds. 1990. *Proceedings of the Michigan morphometrics workshop.* Special Publication No. 2. Ann Arbor, Mich.: University of Michigan Museum of Zoology.

Rouch, R. 1977. Considérations sur l'écosystème karstique. *Compte Rendus Academie Sciences, Paris* 284:1101–1103.

——— 1986. Sur l'ecologie des eaux souterraines dans le karst. *Stygologia* 2:352–398.

Rouch, R., and D. Danielopol. 1987. L'origine de la faune aquatique souterraine, entre le paradigme du refuge et le modèle de la colonisation active. *Stygologia* 3:345–372.

Sarbu, S., T. C. Kane, and D. C. Culver. 1993. Genetic structure and morphological differentiation: *Gammarus minus* (Amphipoda: Gammaridae) in Virginia. *American Midland Naturalist* 129:145–152.

Sbordoni, V. 1982. Advances in speciation of cave animals. In *Mechanisms of speciation,* ed. C. Barigozzi, pp. 219–240. New York: Alan R. Liss.

Sbordoni, V., A. Caccone, E. de Matthaeis, and M. Cobolli-Sbordoni. 1980. Biochemical divergence between cavernicolous and marine Sphaeromidae and the Mediterranean salinity crisis. *Experientia* 36:48–50.
Schluter, D. 1988. Estimating the form of natural selection on a quantitative trait. *Evolution* 42:849–861.
Schmidt, V. A. 1982. Magnetostratigraphy of sediments in Mammoth Cave, Kentucky. *Science* 217:827–829.
Shaw, R. G. 1987. Maximum-likelihood approaches applied to quantitative genetics of natural populations. *Evolution* 41:812–826.
Shoemaker, C. R. 1940. Notes on the amphipod *Gammarus minus* Say and description of a new variety *Gammarus minus* var. *tenuipes*. *Journal of the Washington Academy of Science* 30:388–394.
Shuster, E., and W. B. White. 1971. Seasonal fluctuations in the chemistry of limestone springs: a possible means for characterizing carbonate aquifers. *Journal of Hydrology* 14:92–128.
Silverman, B. W. 1985. Some aspects of spline smoothing approach to nonparametric regression curve fitting. *Journal of the Royal Statistical Society,* ser. B, 47:1–52.
Sket, B. 1965. Taksonomska problematika vrste *Asellus aquaticus* (L.) Rac. (Crustacea, Isopoda), s posebnim ozirom na populacije v Sloveniji. *Razprave Slovenska Akademija Znanosti in Umetnosti* 8:179–221.
——— 1985. Why all cave animals don't look alike: a discussion of the adaptive value of reduction processes. *National Speleological Society Bulletin* 47:78–85.
Slatkin, M, and N. H. Barton. 1989. A comparison of three indirect methods for estimating average levels of gene flow. *Evolution* 43:1349–1368.
Sober, E. 1984. *The nature of selection.* Cambridge, Mass.: MIT Press.
Stearns, S. C. 1980. A new view of life-history evolution. *Oikos* 35:266–281.
Steele, V. J. 1984. Morphology and ultrastructure of the organ of Bellonci in the marine amphipod *Gammarus setosus*. *Journal of Morphology* 181:97–131.
Steele, V. J., and D. H. Steele. 1986. The influence of photoperiod on the timing of reproductive cycles in *Gammarus* species (Crustacea, Amphipoda). *American Zoologist* 26:459–467.
Stock, J. 1986. Amphipoda: Gammarid grouping (Gammaridae s. str. sensu Bousfield). In *Stygofauna mundi*, ed. L. Botosaneanu, pp. 497–503. Leiden: E. J. Brill.
Sulloway, F. J. 1982. Darwin and his finches: the evolution of a legend. *Journal of the History of Biology* 15:1–53.
Swofford, D. L. 1982. *Genetic variability, population differentiation, and biochemi-*

cal relationships in the family Amblyopsidae. M.S. thesis, Eastern Kentucky University, Richmond, Ky.

——— 1991. *PAUP: Phylogenetic Analysis Using Parsimony, Version 3.1*. Computer program distributed by the Illinois Natural History Survey, Champaign, Illinois.

Swofford, D. L., and R. B. Selander. 1981. BIOSYS-1, a Fortran program for the comprehensive analysis of electrophoretic data in population genetics and systematics. *Journal of Heredity* 72:281–283.

Templeton, A. R. 1989. The meaning of species and speciation: a genetic perspective. In *Speciation and its consequences,* ed. D. Otte and J. A. Endler, pp. 3–27. Sunderland, Mass.: Sinauer.

Thines, G. 1969. *L'évolution régressive des Poissons cavernicoles et abyssaux*. Paris: Masson.

Turelli, M. 1988. Phenotypic evolution, constant covariances, and the maintenance of additive variance. *Evolution* 42:1342–1347.

Vandel, A. 1964. *Biospéologie: La biologie des animaux cavernicoles*. Paris: Gauthier Villars.

Vawter, A. T., D. W. Fong, and D. C. Culver. 1987. Negative phototaxis in surface and cave populations of the amphipod *Gammarus minus*. *Stygologia* 3:83–88.

Voneida, T. J., and S. E. Fish. 1984. CNS changes related to the reduction of visual input in a naturally blind fish *Anophtichthys hubbsi*. *American Zoologist* 24:775–782.

Wake, D. B. 1991. Homoplasy: the result of natural selection, or evidence of design limitations? *American Naturalist* 138:543–567.

Wallis, P. M. 1981. The uptake of dissolved organic matter in groundwater by stream sediments—a case study. In *Perspectives in running water ecology,* ed. M. A. Lock and D. D. Williams, pp. 97–111. New York: Plenum Press.

Ward, P. I. 1988. Sexual selection, natural selection, and body size in *Gammarus pulex* (Amphipoda). *American Naturalist* 131:348–359.

Waters, T. F., and J. C. Hokenstrom. 1980. Annual production and drift of the stream amphipod *Gammarus pseudolimnaeus* in Valley Creek, Minnesota. *Limnology and Oceanography* 25:700–710.

Weir, B. S. 1990. *Genetic data analysis*. Sunderland, Mass.: Sinauer.

Welton, J. S. 1979. Life-history and production of the amphipod *Gammarus pulex* in a Dorset chalk stream. *Freshwater Biology* 9:263–275.

White, W. B. 1988. *Geomorphology and hydrology of karst terrains*. New York: Oxford University Press.

White, W. B., and E. L. White. 1991. Karst erosion surfaces in the Appalachian highlands. In *Appalachian karst,* ed. E. H. Kastning and K. M. Kastning, pp. 1–10. Huntsville, Ala.: National Speleological Society.

Wilkens, H. 1971. Genetic interpretation of regressive evolutionary processes: studies on hybrid eyes of two *Astyanax* cave populations (Characidae, Pisces). *Evolution* 25:530–544.

―――― 1973. Über das phylogenetische Alter von Höhlentieren. *Zeitschrift für zoologische Systematik und Evolutionsforschung* 11:49–60.

―――― 1986. The tempo of regressive evolution: studies of eye reduction in stygobiont fishes and decapod crustaceans of the Gulf Coast and West Atlantic region. *Stygologia* 2:130–143.

―――― 1988. Evolution and genetics of epigean and cave *Astyanax fasciatus* (Characidae, Pisces). Support for the neutral mutation theory. *Evolutionary Biology* 23:271–367.

Wilkens, H., and K. Huppop. 1986. Sympatric speciation in cave fish? Studies on a mixed population of epi- and hypogean Astyanax (Characidae, Pisces). *Zeitschrift für zooligische Systematik und Evolutionsforschung* 24:223–230.

Williams, J. G. K., A. R. Kubelik, K. J. Livak, J. A. Rafalsky, and S. V. Tingey. 1990. DNA polymorphisms amplified by arbitrary primers are useful as genetic markers. *Nucleic Acids Research* 18:6531–6535.

Woods, L. P., and R. F. Inger. 1957. The cave, spring and swamp fishes of the family Amblyopsidae of central and eastern United States. *American Midland Naturalist* 58:232–256.

Wright, D. A. 1980. Calcium balance in premoult and post-moult *Gammarus pulex* (Amphipoda). *Freshwater Biology* 10:571–579.

Wright, S. 1978. *Evolution and genetics of populations.* Vol. 4: *Variability within and among populations.* Chicago: University of Chicago Press.

Zeit, L. B. 1993. *The effect of temperature and water level variations on the distribution and abundance of aquatic invertebrates in two cave streams and their resurgence.* M.S. thesis, American University, Washington, D.C.

▼ Index

Acceleration and retardation, law of, 11
Accubogammarus, 71
Adaptation, 26, 29, 32, 46, 123, 132, 165, 187–195
Agassiz, Louis, 8–9
Age: of cave fauna, 18; of caves, 171–174, 185; of invasion, 172, 174–177, 192; of isolation, 174–178
Alkaline water. *See* Hard water
Alleles, 87–89, 92–99, 138, 166, 177–178
Allometry, 25, 106–107, 109
Allozymes, 26, 102, 154, 169, 172, 182
Amblyopsidae, 77, 163. *See also* Fish
Amblyopsis spelaea, 7, 9
Ammonites, 10–11
Amphipods: 2–3, 16, 32–33, 53–57, 60, 62–65, 70, 74, 130, 156, 167; taxonomy of, 168
Amplexus, 124–125, 134, 146–147, 160
Anagenesis, 165
Anophtichthys, 35. *See also* Astyanax
Antennae: 36, 40, 118, 120, 122–130, 136–137; elongated, 26–27, 33, 37, 39, 46, 73, 119; length of, 40, 46, 124–126, 132, 146, 169, 179–181, 188, 190; small, 167
Apomorphic characteristics, 24, 167
Appendages, 4–5, 15, 18, 21, 24, 33, 35–36
Artificial selection, 120
Asellus aquaticus, 15, 34–35, 72, 168. *See also* Isopods
Astyanax fasciatus, 35, 128, 129, 167, 177–178, 180, 185. *See also* Fish

Baetis, 55
Beetles, 8, 33, 156, 168
Behavior, 24, 38, 46, 187
Biotic: environment, 52–54; interactions, 60–65
Biogeography, 70–78
Biplot, 118, 139–146, 165
Blind animals, 4, 7–9, 12, 29, 34, 35
Body length, 37, 46, 56–57, 59
Body shape, 1, 4, 15, 23, 37
Body size, 39, 64, 110, 119, 122, 124, 129–130, 146–147
Brachycentrus, 55
Brain size, 129, 189
Breeding. *See* Reproduction

Caecidotea: holsingeri, 54, 62, 130, 159; *nickajackensis,* 15; *scrupulosus,* 54; *stygia,* 15. *See also* Isopods
Caecobarbus, 163. *See also* Fish
Calceoli, 56
Calcium levels. *See* Hard water
Cambarus bartonii, 64
Cambarus nerterius, 64
Cave-associated traits, 17
Cave fauna, 13–14, 18, 51–52
Cave fish, 16, 20, 72
Cave organisms, 22, 32–35
Caves: environment of, 102; invasion of, 155–162; isolation in, 155–162
Cells, 17
Cepaea nemoralis, 26
Cephalopods, 10
Character reversal, 24–25. *See also* Homoplasy

Index

Circadian rhythms, 24
Cladistics, 2, 27, 33, 167
Cladogenesis, 165
Clustering, 110–114, 116, 139–143, 145
Collembola, 19, 21, 22, 24, 166, 168
Color, 4, 6, 26, 37, 38, 41, 73–74
Compensation, 9, 13, 15
Complexity, 10–11
Compound eye. *See* Ommatidia number
Conductivity, 49, 50, 66–67
Conduit flow, 49–50, 79, 80, 81, 160, 164, 173
Convergence, 163, 164, 167
Cottus: bairdi, 55, 64; *carolinensis*, 130. *See also* Fish
Crangonyx gracilis, 16, 54, 72, 77. *See also* Amphipods
Crickets, 16

Degeneration, 9, 13, 15
Dendogram, 90, 91, 111–112, 115
Diffuse flow, 49, 52
Distribution: of carbonate rock, 73; of *Gammarus minus*, 73
Dixa, 55
Dollo's law, 181–185
Drosophila, 33, 176

Egg quantity, 24, 59, 60, 146–147
Electrophoretic data, 88–89, 100, 101, 111, 112, 114, 115
Electrophoretic distance, 112, 115, 150–155, 168–169, 175
Electrophoretic studies, 86, 102, 104, 116, 167, 183
Endogenous rhythms, 58
Entomobryidae, 19
Eurycea lucifuga, 55
Evolution: of cave organisms, 9, 13, 25–29; morphologic, 15, 118, 179–180; orthogenetic, 10, 12–13; phenotypic, 121; regressive, 25, 35; of troglomorphy, 22–25, 162–165, 172, 191
Evolutionary history, 2, 69–72, 143, 169
Evolutionary rates, 180
Evolutionary theorists, 6–22, 121
Expression point, 12
Eyelessness, 6–7, 26, 71

Eyes: anatomy of, 107, 109, 120, 123, 137; area of, 104–106, 108, 119–120, 122–124, 132, 136–137, 179; degeneration of, 14–16, 18, 20–29, 33–38, 73, 103, 107, 113, 118, 182, 185, 189; loss of, 3, 8, 11–12, 14–16, 28, 33, 70–71, 132, 139, 146, 163, 171, 177–178; morphology of, 38, 40–43, 109, 182; and ommatidia, 38–39, 103–112, 114, 115, 119–120, 122–124, 132, 136–137, 179, 180, 182; size of, 4, 7, 12, 27, 33, 46, 116–117, 125–126, 128–129, 146, 162, 167, 169, 178, 185; statistics on, 120, 124–128

Fecundity, 24, 61, 123–125, 130, 132, 134
Feeding rate, 63
Fish, 16, 20, 24, 39, 64, 72, 128–129, 163, 168, 171, 185, 191
Fitness, 123–125, 133–134, 136, 138–139, 146
Fontigens, 54, 128
Food: and predators, 64–65, 130, 161, 166; and bacteria, 61–63, 67; detritus as, 58–67; feces as, 64; sources of, 21, 63, 124, 162; starvation, 24, 119
Fragility of cave fauna, 4, 23, 37, 38, 119
Full-sibs, 121–122, 146

Galápagos finches, 3, 192–195
Gammarus: archerondytes, 72; *bousfieldi*, 72; *desperatus*, 72; *fasciatus*, 72, 176; *hyalleloides*, 72; *lacustris*, 72; *lawrencianus*, 58; *pecos*, 72; *pseudolimnaeus*, 56, 64, 72; *pulex*, 64, 71; *setosus*, 58; *tenuipes*, 36–39, 46, 74, 119, 166; *troglophilus*, 72. *See also* Amphipods
Ganglion, 42–44
Genetic correlation, 136–147
Geographic setting, 85–86, 104–110
Gnathopods, 36, 37
Gravels, 56
Guano piles, 23
Gyrinophilus porphyriticus, 52, 55, 64

Haeckel's biogenetic law, 10
Haliplus, 55

Index

Hard water, 49–50, 65–67
Harpacticoida, 72
Hawaiian *Drosophila*, 3
Head: length of, 60, 104–106, 108–109, 119–120, 122–129, 131–132, 179; lobes of, 36–37; size of, 8, 40, 41, 105, 106, 107, 120
Heritability, 28, 118–123, 162, 190
Heterozygosity, 87, 103, 160, 171
Homoplasy, 24–25, 27–28, 39, 167, 192
Hyalella azteca, 64. See also Amphipods
Hydrobiid snails, 54, 128
Hydrograph data, 53
Hypertrophy, 23
Hyporheic, 21

Idotea, 15. See also Isopods
Immigration, 9, 13, 15, 17
Inheritance, 9, 13, 15
Insurgences. See Swallets
Ionic balance, 65–66
Isolation species, 155–162, 166
Isopods, 16, 21, 34, 53–54, 63, 130, 156

Karst basins, 78–82
Karst windows: 79–81; populations in, 100–101, 104, 110, 112, 113, 115–116, 130, 151, 167, 168, 172, 181–185

Lamarckism, 6, 9–14, 16–19, 29, 167
Lateral lines, 5, 163
Lava-tube caves, 158
Lepidostoma, 55
Lepomis cyanellus, 55, 64. See also Fish
Leuctra, 55
Life cycle, 33, 54–60, 67, 126
Life span, 24, 29, 33
Light: absence of, 30, 33, 44, 119, 143, 159, 162, 189, 190; levels, 17, 21, 23, 38, 39, 51, 67, 167, 181, 191
Longevity, 6, 29
Lotic environments, 48
Lymnaea, 54

Mandibles, 36
Macrocotyla hoffmasteri, 54
Macroscopic fauna, 54
Matrix correlation, 149–155

Metabolism, 65–66
Metacrangonyctidae, 167
Metohia, 71
Migration rate, 87
Model organism, 34–35
Molanna, 55
Mollusca, 54
Molting, 50, 56, 65, 67
Morphology, 8, 114, 118; change in, 15, 29, 38, 69, 73–75, 188, 194; characteristics of, 24, 35–44, 70, 110, 180; morphological distance, 151–155, 168–169
Morphometrics, 27, 33
Mutation, 25, 177–181, 185

Nautiloids, 10, 11
Neaphaenops telkampfi, 156. See also Beetles
Necturus, 5
Neo-Darwinism, 2, 19–22, 25, 29, 132, 156, 171
Neo-Lamarckism, 9–16, 29, 32, 129, 155, 167, 171
Neoteny, 6, 13
Neotoma, 7
Nervous system, 38, 42, 129, 189–190
Neutral mutation, 27–29
Niphargus, 71–72. See also Amphipods

Olfactory lobes, 39, 42–43, 45, 46, 124
Oligochaetes, 62
Ommatidia number, 38–39, 103–112, 114, 115, 119–120, 122–124, 132, 136–137, 179, 180, 182. See also Eyes
Ontogeny, 11, 122
Onychiurus armatus, 21, 22
Optic elements, 38, 42–43, 44, 45, 106–109, 114, 115, 124
Optic ganglion, 42, 44, 109
Optics, 38, 46, 124, 146, 162, 163, 188
Orconectes australis australis, 6, 72, 77
Organ fragility, 4, 23, 38
Orthogenesis, 10, 12–14, 18
Ostracoda, 72

Packard, Alpheus, 11, 13–16, 29, 129
Parallelism, 19, 25, 29, 163–164, 167

Parent-offspring regression, 121
Path analysis, 149–155, 169
Pereopods, 36, 37, 38
pH, 49–50, 50, 67, 73, 130
Phagocata gracilis, 54
Phenotypes, 124–125, 128, 132–133, 160
Pheromones, 56, 190
Photophobia, 23, 38, 46
Phyletic lines, 10
Phylogenetic state, 2, 10, 13, 24, 26, 33, 166, 187, 191–192
Physa, 54
Physiology, 24, 65–66
Pigment bundles. *See* Retinular cells
Pigment loss, 2–4, 6, 8, 18, 20, 23, 33, 71, 171
Pigment reduction, 24, 26, 29, 35
Pineal gland, 35
Platyhelminthes, 54–55
Pleiotropy, 25, 138, 181
Pleopods, 36
Plethodontid salamander, 13
Poljes, 4
Polymorphic loci, 87
Population: 70–72, 141, 165; density, 49, 54–65, 74, 77, 78, 190; dynamics, 54–60; size, 55, 87, 90, 156; structure, 26, 58, 85, 101, 103, 116, 169, 190–191
Proasellus: albigensis, 21; *vandeli*, 21. *See also* Isopods
Proteolepas, 8
Proteus anguinus, 4–6, 33
Pseudanophthalmus, 33, 168. *See also* Beetles
Pseudosinella, 19, 20; *christianseni*, 22; *hirsuta*, 166, 168

Racial senescence, 18
Recapitulation, 11
Refugium, 156–158, 162
Regressed characteristics, 19, 25
Regressive evolution, 25, 35
Reproduction, 56–61, 67, 118–119, 124, 126, 130, 132, 134–135, 138, 143, 161, 162, 165, 190; cycles, 23, 67; low rates of, 6, 29, 33

Resurgence populations, 104, 108–110, 115–116, 119–120, 122, 130, 147, 151, 163–166, 168
Resurgences (springs), 45, 48–52, 57, 61, 62, 65, 70, 72, 73, 125–128, 159
Retardation, 11–13
Retinular cells, 107–109

Salmo, 64. *See also* Fish
Salvelinus, 64. *See also* Fish
Selection: directional, 123–131, 132, 134–135, 138–139, 146–147, 162, 169, 172, 188–189, 192–193, 195; ecological, 189–190; gradient, 127; natural, 14, 25–27, 32, 45, 102, 117–147, 149, 169, 192; non-linear, 132–136, 147
Semotilus atromaculatus, 64
Sensory organs, 19, 23–24, 129
Sensory papillae, 18
Simplification, 10
Simulium, 55
Snails, 168
Spalax, 6
Speciation, 16, 20, 158
Splines, 133–136
Springs. *See* Resurgences
Statistics, 85–102, 119–147, 160
Structure loss, 7, 16
Stygobromus: 71, 72; *emarginatus*, 33, 54, 62, 64, 159; *spinatus*, 54, 62, 159. *See also* Amphipods
Stylodrilus beattei, 62. *See also* Oligochaetes
Supraesophageal ganglion, 42–43
Swallets, 80, 81

Taeniopteryx, 55
Taxonomic studies, 32–33, 70–72
Temperature: fluctuations, 23, 48–49, 51–52, 58, 67, 70, 72, 77, 130, 156; high, 70, 119, 159; low 48, 52, 66, 70, 159, 163; sensitivity, 23
Terminal addition, 11
Traits, 83, 106, 119, 120, 123, 126, 132, 136, 161
Troglobites, 158
Troglomorphy: 29, 33; characteristics of, 23, 32, 34, 37–38, 46, 47, 130; evo-

lution of, 17, 22–25, 162–165, 172; syndrome, 25, 32; traits of, 23, 24, 35, 65, 77, 118–119, 124, 151
Troglophiles, 158
Trogloxenes, 158
Turbulent flow, 79
Typhlichthys, 163. *See also* Fish

Uropods, 36–38, 167
Urosome, 37, 128
Use and disuse. *See* Lamarckism

Wings, 8, 24

Zenkevitchia, 71

Bei Fragen zur Produktsicherheit wenden Sie sich bitte an:
If you have any questions regarding product safety,
please contact:

Walter de Gruyter GmbH
Genthiner Straße 13
10785 Berlin
productsafety@degruyterbrill.com